EARLY
ELECTRODYNAMICS
THE FIRST LAW OF CIRCULATION

R. A. R. Tricker

PERGAMON PRESS

OXFORD · LONDON · EDINBURGH · NEW YORK
PARIS · FRANKFURT

Pergamon Press Ltd., Headington Hill Hall, Oxford
4 & 5 Fitzroy Square, London W.1

Pergamon Press (Scotland) Ltd., 2 & 3 Teviot Place, Edinburgh 1

Pergamon Press Inc., 122 East 55th Street, New York 22, N.Y.

Pergamon Press G.m.b.H. Kaiserstrasse 75, Frankfurt-am-Main

Federal Publications Ltd., Times House, River Valley Road, Singapore

Samcax Book Services Ltd., Queensway, P.O. Box 2720, Nairobi, Kenya

First Edition 1965

Copyright © 1965
Pergamon Press Ltd.

Library of Congress Catalog Card No. 65-19838

Set in 10/12pt. Times
and Printed in Great Britain by
James Upton Limited, Birmingham.

Contents

Preface

A T the beginning of the year 1820 nothing was known of the magnetic action of an electric current. By 1826 the theory for steady currents had been completely worked out, Since then, though newer methods may have made the handling of the mathematical apparatus simpler and more concise, nothing fundamental has been changed. The object of this book is to trace this branch of electrical theory to its origins and to show how the results achieved in those few early years have led to the various theorems which have since been deduced. This is not to maintain that the basis for the application of the theory is still to be found in the comparatively crude experiments of the early nineteenth century but an appreciation of its origin can lead to a better understanding of the logical structure of the theory.

The manuscript was read by Dr. Mary Hesse who made several suggestions of which the author availed himself and he is similarly indebted to Dr. ter Haar, the editor of the series, for advice during the preparation of the book. He would also like to thank the Editor of the *School Science Review* for permission to reproduce figures from some of his articles in that journal, his former colleague Mr. R. C. Lyness for many discussions of various points and his wife for her help in the preparation of the manuscript.

The author's thanks are also due to the Danske Tekniske Höjskole for the portrait of Oersted and to the commune de Poleymieux au Mont.d'Or who provided the portrait of Ampère.

Cropston, 1965

R.A.R.T.

Part 1

Part 1

1
The Stage

THE developments with which this book is concerned all occurred within a very short period of time, beginning in 1820 and virtually complete in 1826, and were centred on Paris, then at the peak of its influence in the scientific world. Here, in this chapter, we must set the stage for a consideration of this rapid advance in knowledge by assembling the views which were held immediately prior to this period.

Contrary to what might perhaps have been expected, the eighteenth century, following the publication of the *Principia* of Newton in 1687, did not lead to a rapid growth throughout the field of science. Instead it proved a period of consolidation. The consequences of the Newtonian theory of mechanics were worked out in great detail by mathematicians such as D'Alembert (1717–83), Lagrange (1736–1813), Laplace (1749–1827) and Legendre (1752–1823). As the names show, this was the work of French mathematical genius. It was a theoretical development accomplished by the employment of principles of the most general application, such as that of D'Alembert (1743). Little was added to the fundamental basis of physical facts, the development taking the form of the construction of an elaborate deductive system on the general lines of Euclid's geometry. The birth of mathematical physics may be looked upon as having occurred with Galileo and Newton, but it reached adult stature, if such a phrase has any meaning, in this period, and it took place mainly in the field of gravitational theory.

Lagrange deduced the Principle of Least Action from Newton's laws of motion in 1760. He first employed the idea, though not the name, of potential in 1773. In 1784 Laplace formulated his famous equation,

$$\frac{\partial^2 V}{\partial x^2} + \frac{\partial^2 V}{\partial y^2} + \frac{\partial^2 V}{\partial z^2} = 0$$

though, in fact, he gave it in the more complicated form in polar coordinates.

As far as electricity was concerned, however, all this equipment remained largely off-stage in the wings. The basic physics of the science had not been sufficiently elucidated for these general principles to be applied to it. The stage had still to be set. This preparation occupied the whole of the eighteenth and the first two decades of the nineteenth centuries. Even as late as the year 1800 ideas on the actual physical phenomena occurring in the field of electricity appear to have been very confused. For the first half of the eighteenth century the general idea of electrification appears to have been of "effluvia" emanating from charged bodies. In that attention was thereby concentrated on the surrounding space rather than on the bodies themselves, this idea possessed some affinity with the field theories of the second half of the nineteenth century. Vortex theories to account for the movement of small light particles attracted to an electrified body were put forward. The presence of these vortices was thought to be apparent to the senses through the tickling sensation experienced when a charged body is brought close to the face. The effluvia were not very clearly distinguished from heat and light.

Effluvia virtually vanished from the scene with the work of Franklin (1706–90), who, in 1747, developed his one-fluid theory of electricity. He had been preceded by Charles-François Dufay, the superintendent of the Royal Gardens in France, who had found, in 1733, repulsion between some charged bodies and attraction between others (notably between glass and copal). From this he concluded that there were two electric fluids and he attached the adjectives vitreous and resinous to them. We shall see that in Paris, even as late as 1820, his notion of two electric fluids still held the field. Franklin's one-fluid theory failed to explain the mutual repulsion of negatively charged bodies, which possess less of the electric fluid but whose particles of ordinary matter would exert gravitational forces only on each other. In this respect the two-fluid theory was superior. A two-fluid theory entails the annihilation of the effects of each fluid by the presence of the other, but in uncharged bodies the two electricities were thought of as combined. The combination was decomposed in the process of charging a body electrically either by friction or other means—including the voltaic pile later on, when it had been invented. Even the single-fluid theory had to assume repulsion between particles of ordinary matter to account for the action of both kinds of charged bodies on those similarly charged. It was, in fact, Aepinus who, denying the idea of effluvia, postulated this repulsion between particles of ordinary matter. He did this in 1759 at St. Petersburg, where he had gone in 1757. Aepinus's theory was thus not altogether unlike the theory of electrons of a century and a half later. Until the law of force was discovered, however, the mathematical development of the science was impossible.

By 1750 the one-fluid theory of Franklin had virtually included the concept of action at a distance. "Though the particles of electrical matter do repel each other, they are strongly attracted by all other matter." Franklin identified his electric fluid with Dufay's vitreous electricity.

The honour of first showing that in electrostatics the law was one of the inverse square of the distance, probably belongs to Priestley, a friend of Franklin. He performed experiments with a hollow charged conductor and demonstrated that there was no charge on the inside. Moreover, from this he drew the correct conclusion, which he obtained from his knowledge of Newton's theory of gravitation, that electric attractions obey the same law as gravitational attractions, and this he published in 1767. The power of the method does not appear to have been sufficiently appreciated at the time.

In 1769 John Robison, of Edinburgh, determined the law of force experimentally. He obtained somewhat different results for attraction and repulsion but conjectured that the correct result was one of inverse squares. It was, however, Coulomb (1736–1806) who is commonly accredited with the first experimental determination of the law of force between two charges. This he did in 1785 by determining the force between charged spheres by means of a torsion balance, an instrument which had already been employed by John Michell, a Fellow of the Queens' College, Cambridge, who by its means, had arrived at a similar law for magnetic poles. Coulomb's findings were generally accepted, to which result their elegance and simplicity in large measure contributed. The charges on two conducting spheres, such as Coulomb used, are not distributed uniformly over their surfaces when placed near each other, and the problem of the actual distribution was not solved for another 30 years—in 1813 to be precise, by Poisson (1781–1840). Electrical leakage is also likely to make the torsion balance a far from accurate instrument in hands less skilled than Coulomb's. However, the mathematical theory of electrostatics was thus equipped with the physical information required for its development, but it had still to wait for Poisson for this to take place. His first Mémoire on the subject was published in the *Mémoires de l'Institut* for 1811, and with it the science of electrostatics changed practically overnight from a rather confused qualitative topic to a mathematical study of considerable complexity. Not only did he borrow results from gravitational theory, such as Laplace's equation, but he introduced the equation known by his own name, for points where the attracting material resides. The mathematical theory of electrostatics was thus available at the beginning of the period with which we shall be concerned in subsequent chapters.

The same is not quite true for the theory of magnetism. Though starting earlier, it was a little later in reaching such a polished state.

The conception of magnetic poles as places from which the magnetic force originates, goes back as far as Peter Peregrinus of Maricourt, who wrote on magnetism in 1269. It was he who first introduced the term "pole" into the science, and he based it on his examination of a lodestone with a search needle, which revealed the presence of points on the surface analogous to the poles of the earth. Gilbert, of Colchester (1540–1603), who is looked upon as the founder of the science of magnetism, first suggested that the earth itself was a magnet. With him magnetic attractions came into prominence, and at one time they were candidates for the place of supporters of Kepler's astronomy.

Newton considered this candidature and rejected it. He also considered the couple exerted by one short magnet on another placed near it and reached the correct result that it would vary inversely as the cube of the distance. The science then made virtually no further progress until 1750 when, as we have already seen, the inverse square law for magnetism was discovered by John Michell with the help of his torsion balance. (Michell became Rector of Thornhill, in Yorkshire, in 1767, and lived there until his death in 1793.) It was with magnetism, too, that Coulomb carried out his first experiments with a torsion balance, by which he also arrived at the law of inverse squares. This he did only shortly before his experiments with electrostatic charges, so that the two sciences of electrostatics, on the one hand, and magnetism, on the other, drew level at this point.

As in electrostatics, a two-fluid theory held sway in magnetism on the Continent, in spite of the attempts by Aepinus to introduce a single-fluid theory on the same lines as that of Franklin for electrostatics. The two-fluid theory was later advocated by Anton Brugmans and by Wilcke. The two fluids received the names austral and boreal, a *north-seeking* pole being a region where the *austral* fluid predominated, and a *south-seeking* pole a region where the *boreal* fluid predominated. This choice was made in view of the fact that a north-seeking pole was homologous with the south pole of the earth, by which it was repelled. The south-seeking pole was similarly homologous with the north pole of the earth, and must therefore be the seat of boreal magnetism.

There was much discussion at the end of the eighteenth century of the close analogy which appeared to exist between magnetism and electrostatics. Some connection was suspected. Lightning had been known to magnetize pieces of steel. In both cases the same law of the inverse second power of the distance appeared to be obeyed, but the actual connection proved too elusive before the discovery and theoretical development of current electricity. There were, moreover, striking and important differences between the two groups of phenomena. Electric charge of either kind could easily be had isolated from the other on a charged body. This was not the case with the magnetic fluids, which turned out to be impossible to separate. A great range of bodies could be electrified; only a few could be magnetized. There was nothing in the field of magnetism to correspond to conductors or insulators in electricity. Thus the particles of the magnetic fluids were looked upon as inseparable from the molecules of matter. In the papers with which we shall be dealing in this book, the expression, a molecule of boreal magnetic fluid and a molecule of austral magnetic fluid or magnetism, occurs frequently. A molecule of boreal magnetism is synonymous with what we now call a south pole, and one of austral magnetism with a north pole.

Poisson performed for magnetism a very similar service to that which he had rendered for electrostatics a few years earlier. In 1824 he published, in the *Mémoires de l'Académie des Sciences de Paris*, a theory of magnetism as advanced as that which he had already contributed for electrostatics. In this, Poisson accepted the two-fluid theory of magnetism, though his results are easily adapted to any theory of magnetic poles. His theory was thus made known to the world only towards the end of the period we shall have under our immediate consideration in the main part of this book, and so probably had not much influence upon it. Ampère refers to his paper in his Mémoire of 1825. No doubt Poisson was working on his theory at the same time as Ampère was pursuing his researches. In this task he would, therefore, not have available the finished product of Ampère's theory of magnetism, so that the two theories thus grew up largely independent of each other. This was a pity, and it goes far to explain the way in which the theory of electricity developed. The result was a compromise and one of a rather curious form. The common practice in electrical theory is for the theory of magnetism to be developed along the lines of Poisson's theory and for the theory of electrodynamics to be constructed on this as a basis. Finally, however, the magnetic pole is dropped, and it is the electric charge which is given pride of place. This is Ampère's point of view. Though it is only a matter of form, it involves pulling up the foundations after the superimposed edifice has been completed, which is not a very elegant method of procedure. As Ampère pointed out, it is only by using the electric current for the foundation that it is possible to unify the two sciences of electricity and magnetism into one conceptual scheme and thus provide a solution to the problem which defeated the physicists of the eighteenth century.

The action which forms the main consideration of this book is centred round the discovery of current electricity. Some tentative steps had already been taken in this direction in the course of the study of electrostatics. Thus the idea of the conservation of charge was embedded in the theories of electric fluids already, as used by Franklin and others. The result was substantiated to some extent by Franklin's experiments concerning the charges carried by persons standing on insulating stools. Thus if a person so insulated charges another similarly insulated by means, say, of a glass rod, the second person can pass a spark to a third standing on the ground. If, however, the first two touch each other after the charging, this power is lost. Conductors and insulators had been distinguished—first as "non-electrics" and "electrics" but later, after it had been found possible to charge metals when they were insulated, as bodies through which the electric fluid could or could not pass. In 1729 Gray and Wheeler had conveyed charges from a glass rod over a distance of 886 ft through a packthread supported by silk loops. In 1775 the Hon. Henry Cavendish had compared the conductivities of various substances by the simple method of estimating the intensity of the electric shock which he received. His results appear of surprising accuracy when compared with present-day measurements.

The science of current electricity is usually looked upon as having originated in 1780 with the famous observations of Galvani (1737–98) on the stimulation of the muscles in the legs of freshly killed frogs. Galvani held the post of Professor of Anatomy in his native city of Bologna. His first observation was of the contraction of the muscle in the leg of a frog when the nerve happened to be touched by a scalpel at the same time as someone nearby drew a spark from an electrical machine. The observation was thus immediately associated with electricity and, in the course of further investigation, it was traced to metallic contacts between the nerve and a circuit of metals. Galvani's ideas, however, were that the electricity originated in the animal and that the metallic arc served merely to discharge it, like the discharging tongs of a Leyden jar. Galvani died in 1798 after having been deprived of his post because he refused allegiance to the new Cis-Alpine Republic set up by Bonaparte.

Volta (1745–1827) took a different view and considered the source of electricity to arise from the contact of different metals. He had already invented the electrophorus in 1775. In 1800 he discovered how to construct the electric pile, which gave much more spectacular results than had before been obtained in the field of current electricity, and which had only been associated with electrostatics. He explained its action on the basis of a contact theory. He showed that discs of copper and zinc, mounted on insulating handles, became charged on being placed in contact. On separation they behaved in a similar manner to the electrophorus or condensing electroscope, and Volta had been able to detect the charge resulting from the contact by means of a sensitive electroscope. He looked upon the electrolyte held in solution in the pasteboard discs placed between the metal discs of the pile as functioning solely as conductors, and regarded the generation of the resulting current when the opposite terminals were connected as something akin to perpetual motion. Though the search for mechanical perpetual motion had by then been abandoned by scientists, this was not thought to preclude such motion in the case of the imponderable fluids operating in the electric current. In this theory the performance of the Leyden jar was an important analogy employed to help in thinking about the problem. Metallic contact was looked upon as providing a continual charge of the pile which was thus something of a widow's cruse. The analogy was helped by the fact that it was possible to feel a shock from a pile just as from a Leyden jar, though probably not so intensely.

The electric pile was set up in England as soon as news of it reached the country, which it did in a letter from Volta to Sir Joseph Banks, President of the Royal Society, in 1800. In this country it was the chemical action associated with electrolysis which attracted most attention and particularly that of Sir Humphry Davy who evolved a theory of the pile of a chemical nature. He emphasized the importance of the substances in solution in the moist pasteboards and their role in "oxydating" the zinc. As a result of these investigations a much more efficient pile, using jars of acid, was evolved, and during the period upon which we shall be concentrating, news of it had reached Paris, where references are to be found to the *pile anglais*. It was not, however, in use there at the time the developments with which we shall be concerned were taking place, although a very large pile had been constructed in the Institut Polytechnique, on the orders of Napoleon. The English pile was able to raise fine wires to incandescence, using only a single pair of plates, a fact with which Ampère was impressed. It makes a sad comment upon modern progress to realize that in 1813 Sir Humphry Davy, accompanied by the youthful Faraday, was able to visit Paris in the middle of the Napoleonic war. Here they met Ampère, Cuvier, Gay-Lussac, Humboldt and many others.

Indeed, they travelled freely across France and into Italy, and they do not appear at any time to have taken particular pains to temper their language to suit the politics of the country in which they found themselves. It is recorded that after a tour of the Louvre, where were collected masterpieces pillaged from half the capitals of Europe, Davy remarked: "What an extraordinary collection of fine frames."

The science of electricity, as it existed in academic circles in France at the time the curtain rises upon our principal actors, is thus characterized by two-fluid theories in both the fields of electrostatics and magnetism. Each of these branches was looked upon, rather reluctantly it is true, as entirely distinct from the other. They presented many similarities between themselves and with gravitational theory, but no connecting link had been found. The contact theory tended to dominate thinking about the electric pile. The difference between electric charge and electric tension was but dimly discerned. Their relation to work and energy had not been elucidated, and the same was true of heat and light. Deduction from Newton's laws had led to the discovery of the conservation of mechanical energy, but the conservation of energy in general and its conversion from one form into another had not been realized. It was in such a climate of opinion that the announcement of Oersted's discovery of the effect of an electric current on a compass needle was made to the scientific world in 1820, and in which was evolved, within a space of six short years which followed, the complete theory of the electrodynamics of steady currents, still employed today. Before the announcement nothing was known of the magnetic action of electric currents. At the end of that period the theory had been given a form which has served as a basis for all subsequent work.

II
Dramatis Personae

HANS CHRISTIAN OERSTED, 1777–1851 (Plate I)

H. C. Oersted was born on 14 August 1777 in Rudkjöbing on the island of Langeland, off the east coast of Denmark. He attended the university of Copenhagen, from which he graduated as a pharmacist in 1797. He was awarded gold medals for essays in aesthetics and medicine and, in 1799, the degree of Ph.D. for a dissertation on Kant's philosophy. He was elected Professor at Copenhagen University in 1806 after extensive travelling abroad.

PLATE I. H. C. Oersted. From a portrait kindly provided by the Danske Tekniske Höjskole.

His most important contribution to science was his discovery of the action of an electric current on a compass needle. Of that there can be no doubt, and it will be that alone with which we shall be concerned in this book. To chemistry he contributed the preparation of metallic aluminium. He devoted much of his researches to a study of the compressibility of liquids and gases.

Oersted's researches in electricity were motivated by his metaphysical belief in the unity of the forces of nature. This, no doubt, he derived from the *Naturphilosophie* current in Germany at the beginning of the nineteenth century, the chief exponent of which was F. W. J. Schelling. This emphasized knowledge *a priori* and deprecated mere empirical knowledge, a view which Kant, whom Oersted had also studied, tended to support. It seems, indeed, a curious doctrine to inspire an experimental scientist. Oersted, trained as an apothecary, was, however, by no means uncritical of this depreciation of empirical knowledge. "Schelling wants to give us a complete philosophical system of physics but without any knowledge of nature except from textbooks."

His attempts to demonstrate the unity between galvanic and magnetic forces, however, were not successful until 1820. An account of what happened on that occasion has been given in a well-known letter to Michael Faraday from Professor Hansteen dated 30 December 1857.

"Professor Oersted (he wrote) was a man of genius but he was a very unhappy experimenter; he could not manipulate his instruments. He must always have an assistant, or one of his auditors, who had easy hands, to arrange the experiment. I have often, in this way, assisted him as his auditor.

"Already in the former century there was a general thought that there was a great conformity, and perhaps identity, between the electrical and magnetical force; it was only a question of how to demonstrate it by experiments. Oersted tried to place the wire of his galvanic battery perpendicular (at right angles to) over the magnetic needle, but marked no sensible motion. Once, after the end of his lecture, as he had used a strong galvanic battery in other experiments he said, 'Let us now once, as the battery is in activity, try to place the wire parallel to the needle'; as this was made, he was quite struck with perplexity by seeing the needle making a great oscillation (almost at right angles with the magnetic meridian). Then he said: 'Let us now invert the direction of the current' and the needle deviated in the contrary direction. Thus the great detection was made; and it has been said, not without reason, that 'he tumbled over it by accident'. He had not before any more idea than any other person that the force should be transversal. But, as Lagrange has said of Newton on a similar occasion, 'Such accidents only meet persons who deserve them'."

This letter was, of course, written after a lapse of 37 years. There is one point in it which calls for comment. If Oersted disposed of currents which were great enough to deflect a compass nearly at right angles to the magnetic meridian, one might have imagined that he would have obtained a reversal of the needle on occasion, when he held the wire at right angles to the meridian. One would have thought, too, that even when the field of the current had the same direction as that of the earth, slight movements of the wire would have produced some movement in the needle which would have made it obvious that it was following the direction at right angles to the wire, rather than the field of the earth. It is possible that Professor Hansteen exaggerated the magnitude of the effect which Oersted was able to produce. In any case his account was not written until nearly 40 years after the event and may not be at all accurate.

Oersted himself has left three accounts of how he made his famous discovery, which agree among themselves but which are at variance with Professor Hansteen's description. An extract from the first of these, of which there are French and German editions, is as follows.

"Since for a long time I had regarded the forces which manifest themselves in electricity as the general forces of nature, I had to derive the magnetic effects from them also. As proof that I accepted this consequence completely, I can cite the following passage from my *Recherches sur l'identité des forces chimiques et électriques*, printed in Paris 1813. 'It must be tested whether electricity in its most latent state has any action on the magnet as such.' I wrote this during a journey so that I could not easily undertake the experiments; not to mention that the way to make them was not at all clear to me at that time, all my attention being applied to the development of a system of chemistry. I still remember that, somewhat inconsistently, I expected the predicted effect particularly from the discharge of a large electric battery and, moreover, only hoped for a weak magnetic effect. Therefore I did not pursue with proper zeal the thoughts I had conceived; I was brought back to them through my lectures on electricity, galvanism and magnetism in the spring of 1820. The auditors were mostly men already considerably advanced in science; so these lectures and the preparatory reflections led me on to deeper investigations than those which are admissible in ordinary lectures. Thus my former conviction of the identity of electrical and magnetic forces developed with new clarity, and I resolved to test my opinion by experiment. The preparations for this were made on a day in which I had to give a lecture the same evening. I therefore showed Canton's experiment on the influence of chemical effects on the magnetic state of iron. I called attention to the variations of the magnetic needle during a thunderstorm, and at the same time I set forth the conjecture that an electric discharge could act on a magnetic needle placed outside the galvanic circuit. I then resolved to make the experiment. Since I expected the greatest effect from a discharge associated with incandescence, I inserted in the circuit a very fine platinum wire above the place where the needle was located. The effect was certainly unmistakable, but it seemed to me so confused that I postponed further investigation to a time when I hoped to have more leisure.† At the beginning of July these experiments were resumed and continued without interruption until I arrived at the results which have been published."

About 2 months after the first announcement, Oersted submitted a second paper to the Académie des Sciences. It was received on 29 September 1820, 25 days after Arago had announced Oersted's discovery. By that time Ampère had already announced—on 25 September—the mutual action of two currents.

JEAN-BAPTISTE BIOT, 1774–1862

J. B. Biot was born in Paris in 1774, was educated at the Lycée Louis le Grand and at the École Polytechnique, and was trained as an artilleryman. He became Professor at the École Centrale at Beauvais, and in the year 1800 he was appointed to the Chair of Physics in the Collège de France, a post he held for 26 years. His interests were many and varied, and by the year 1820, when our story commences, he had attained a position of seniority. He produced a number of mathematical works, including an analysis of Laplace's celestial mechanics, and a treatise on curves and surfaces of the second degree, but it was as an experimentalist on which his claims to fame rest. His *Traité Elementaire de Physique* and his *Précis Elementaire de Physique* became standard textbooks. He went on several expeditions to measure variations in gravity, to

---

†"All my auditors are witnesses that I mentioned the result of the experiment beforehand. The discovery was therefore not made by accident, as Professor Gilbert has wished to conclude from the expressions I used in my first announcement."

examine meteorites and to measure the length of a degree along the meridian of Paris. He was a member of the Bureau des Longitudes. He investigated the properties of polarized light and wrote on the ancient history of astronomy. His interests and activities were maintained throughout a long life. He died in 1862.

In 1804 he accompanied Gay-Lussac on his first ascent by balloon. A little earlier in the year the Academy of Saint Petersburg had asked Robertson and Sacharoff to make an ascent to ascertain whether or not the earth's magnetic field varied appreciably with height. The Institut de France felt that they should try to carry out a similar investigation for themselves, and assigned the task to Gay-Lussac and Biot. Laplace obtained for them, from M. Chaptal, the Minister of the Interior, a small balloon which had been used in Napoleon's expedition into Egypt. In the basket beneath this they ascended from the gardens of the Conservatoire des Arts et Métiers, at 10 o'clock in the morning of 24 August 1804, equipped with the apparatus with which they hoped to carry out the measurements. At a height of about 4000 ft they broke through the clouds which throughout their trip lay like a sea of foam beneath them. At 9000 ft they released a bee and at 11,000 ft a finch. The latter perched for an instant in the rigging and then plunged downwards as though acted on only by gravity. At a height of about 13,000 ft they commenced the experiments which were the object of the whole expedition; they attempted to time the oscillations of a small magnet suspended horizontally. However, their plans had not envisaged the possibility of the rotation of the balloon, and the experiment, simple enough in theory, proved very difficult to carry out in practice. They succeeded in timing five oscillations of the needle, but the results were not thought to be satisfactory. Apart from the light it throws upon the character of Biot in being ready to attempt such an expedition, the story is interesting from our point of view in that it shows that, as early as 1804, he was measuring magnetic fields by the same method which he employed in 1820 in his investigation of the magnetic action of electric currents, and, indeed, he must already have been looked upon as one of the experts in this field to be chosen to make the ascent.

It is of interest to notice, in passing, that Gay-Lussac later repeated the ascent, this time by himself. He then attained a height of 23,000 ft.

Arago tells the story of how, on that occasion, having approached this height, Gay-Lussac wished to ascend even higher and threw overboard everything with which he could possibly dispense. Among the articles thus jettisoned was a chair in white wood which landed in a thicket near a country-girl minding her sheep. The sky was clear, the balloon invisible. What else could the chair be but a piece of the furniture fallen out of heaven? The only argument against such a view was that the craftmanship was crude and the heavenly craftsmen could hardly be so inept. There, indeed, the matter had to remain until the news of the flight of Gay-Lussac was published abroad.

FÉLIX SAVART, 1791–1841

Félix Savart was born in 1791 in Mézières, the son of the director of the artillery workshops. He took up a medical career. In 1819 he submitted a mémoire to Biot, who took an interest in the young man and encouraged him to go on with his researches. He procured for him a post as Professeur de Physique which he held for seven years, and it was during this time that he collaborated with Biot in the experiments on the magnetic field of electric currents, with which his name is now linked.

Later, in 1827, he was elected a member of the Académie and in 1828 became Conservateur du Cabinet de Physique at the Collège de France, where he was nominated Professor of Experimental Physics, following Ampère. His published papers are mainly on oscillations and sound, the human voice and the voice of birds.

ANDRÉ-MARIE AMPÈRE, 1775–1836 (Plate II)

André-Marie Ampère is the principal actor in the early development of electrodynamics, a word which he coined himself. He attained the highest honours which science had to offer, was a member of the chief scientific societies all over the world, and became universally acclaimed by his contemporaries. Yet in spite of this success, in his private life he presents a truly tragic figure.

Towards the end of his life he confessed that, of the whole, only two years had brought him any real happiness.

André - Marie AMPÈRE

1775 - 1836

PLATE II.   From a portrait kindly provided by Monsieur le Maire, Poleymieux au Mont.d'Or.

He was born in Lyon on 20 June 1775. His father was a retired merchant living at Polémieux-lez-mont-d'Or where André-Marie spent his infancy. There was no school in the village and Ampère was entirely self-taught. In early life he liked Buffon's *Histoire Naturelle* to be read to him. His father started to teach him Latin but, seeing the boy had an aptitude for mathematics, he postponed the study of Latin and left his son to follow his own desires, making himself responsible for providing him with books. The story is told of how, on one occasion, Ampère, at the age of 11, was taken by his father to the library of the Collège de Lyon, where he surprised the Abbé Daburon, the librarian, by asking for books by Euler and Bernoulli. On being told that they were among the most difficult ever written, he replied that nevertheless he thought he would be able to study them. On finding that those of Bernoulli were in Latin he quickly mastered sufficient of that language to be able to read them. At 18 he had started the *Mécanique Analytique* of Lagrange and made nearly all the calculations in it again, for himself. The twenty volumes of the encyclopedia of Diderot and d'Alembert he knew from A to Z. Fifteen years later he could recite whole passages from it by heart.

Then the storm broke. His father had become justice of the peace at Lyon and president of the Tribunal Correctional. However, he took part in the revolt of the city against the Convention. When the city had been retaken, he was arrested and guillotined. The evening before his execution he wrote to his wife regretting that he was unable to leave her even comfortably off. His greatest expense, he said, had been the purchase of books and geometrical instruments for their son, which he thought had been a wise economy, since the boy had had no other master but himself.

The effect of this calamity on the young André-Marie was overwhelming. He retired into himself and stared vacantly at the sky or made little heaps in the sand in front of him. He became a near idiot, and the effect lasted, in this intense phase, for a whole year. Beyond doubt his whole life was affected by it. At the end of this year he began to be revived by contact with J. J. Rousseau's *Lettres sur la botanique*. There then followed a phase of poetry in which he wrote large quantities of curious verse. It was in the fields, on a botanical excursion, that he met the lady whom he married on

2 August 1799. She was Julie Carron, not rich but deeply religious, an attitude she succeeded in imparting to her husband. Intense religious feeling, punctuated by periods of tormenting doubt, marked the whole of Ampère's later life. He himself recorded that there were three events which altered the course of his life. They were: his first communion, his reading of Thomas's *Eloge* on Descartes, and the Fall of the Bastille.

At 24 he had two happy years with his wife before him. A son, Jean-Jaques, was born to them on 12 August 1800. The son later became well known on his own account, as well as adding to his father's worries at other times; he wrote on poetry, on medieval French literature and on the foundations of the French language, ultimately becoming Professor at the Collège de France. Ampère lived with his family at Lyon and sought a post by which he could augment his meagre means. He could not obtain one in Lyon but was appointed to the Chair of Physics and Chemistry at the École Centrale of the Department of Ain, in December 1801, and this entailed his transference to Bourg. His wife was ill and unable to accompany him. She died in 1803. Ampère was profoundly affected by his wife's death, which led to a further crisis of religious fervour.

While at Bourg he wrote his first mathematical paper on the theory of probability, and this was favourably received by Lalande and Deslambre. Ampère was, in consequence, offered the post of Professor at the Lycée de Lyon, a post which it had been his ambition to achieve.

He married for a second time, largely on the advice of friends, but it was not a happy marriage. His second wife did not wish for children at any price, and, when she knew that she was to bear one, she made life intolerable for him. He learnt of the birth of his daughter, Albine, from the concierge. The marriage broke up on 11 January 1808. The second Mme Ampère lived until 1866 and died at Versailles.

Academic honours now showered upon him. In 1805 he was Répétiteur d'Analyse at the École Polytechnique. In 1806 he was elected member of the Bureau Consultative des Arts et Métiers. In 1808 he became Inspecteur Général of the University. In 1809 he was Honorary Professor at the École Polytechnique and Chevalier of the Légion d'Honneur. By 1814 he was a member of most of the learned societies in Europe, including the Royal Society of London, in spite of the state of war between England and France. Arago recalls that for many weeks scientists from France and from abroad would crowd his humble laboratory in order to see demonstrated the orientation in the earth's magnetic field of a piece of platinum wire carrying an electric current.

In addition to his contributions to mathematics and physics, Ampère also cultivated an interest in ancient literature and history. He wrote extensively in philosophy, where he tried to reconcile religion with reason. He maintained the certainty of metaphysics. His last major work was on the classification of the sciences, in which he was able to employ his great breadth of knowledge.

Ampère possessed great candour combined with extreme sensitivity, simplicity and timidity. Throughout his whole life he felt himself drawn to his old home-town of Lyon where he had left his first and best friends. He was most inept in the management of his own affairs. He was never rich, yet he spent large sums on his scientific apparatus and gave away more than he could afford. Neither his son nor his daughter proved free of worry for him. Having just had a piece accepted by the Odéon his son abandoned it all to follow a Mme Recamier to Italy. His daughter married an officer, a gambler crippled with debt and an alcoholic. In addition Ampère tended to take upon his own shoulders the cares of the world, and they were many in the France of 1815. In that year he wrote to his friends in Lyon: "I am like grain between two millstones. Nothing could express the heart rendings I experience and I have no longer the strength to support life."

He suffered with severe myopia and read with difficulty. Once he met a young man similarly afflicted who lent him his spectacles, and for the first time he saw the marvels of light on the hills, a miracle which has to be experienced to be appreciated.

He never lost the habit of retiring within himself, especially during periods when he was troubled with religious doubts, and giving himself up to meditation. He came to be looked upon as a very bizarre character and accumulated round himself stories of the absent-minded professor. Once it starts there is no end to such a process. Most of these stories were doubtless purely apocryphal. They are like filings of soft iron which, though not magnets themselves, adhere to any magnet to which they are brought near. It was of Ampère that the story was first told of how, on the way to deliver a lecture, he noticed a peculiarly marked flint near the Pont des Arts, in Paris. He was examining this when it occurred to him that he would be late for his lecture. He took out his watch to see the time and then strode on, putting the stone in his pocket and throwing his watch in the Seine. In his lectures he was prone to mop his brow with the blackboard duster and to wipe the board with his handkerchief.

On another occasion, on his way to lecture, he was said to be turning over some problem in his mind and sought a surface on which he could put down some algebra which even he could not carry in his head. Seeing the black back of a cab drawn up by the kerb, he took out a piece of chalk from his pocket and soon had the surface covered with hieroglyphics. He was just on the point of reaching a solution when, to his consternation, the carriage drove away carrying his work with it.

Such is the mixture of triumph and sadness presented by the life of the great Ampère. He died in 1836. He had left for Marseilles on an inspection of the University, already suffering from an acute infection, and arrived there at the end of his strength. On 9 June, as he lay on his pillow, the Proviseur of the College commenced to read to him passages with hushed voice. Characteristically, Ampère stopped him and declared he knew the book by heart. He died on the following day.

III
Commentary†

A. THE WORK OF OERSTED

In this section we shall examine the papers, written in the early nineteenth century, which established the rules according to which the forces which one circuit carrying an electric current exerts upon another nearby, may be calculated. The work on which these conclusions immediately rested, was accomplished, as we have seen, in the space of 6 years at most. This development was not, however, quite as straightforward as the rapid development of electrostatics, which took place rather earlier, at the beginning of the nineteenth century and the last part of the eighteenth, which we have already mentioned. This came about largely through the direct application of the results of the gravitational theory of attraction to the analogous case of electrostatics. The development of the theory of current electricity, on the other hand, required the establishment of entirely novel concepts and was an achievement requiring an intellect of the first order.

Nevertheless, the work of Newton and his successors, such as Lagrange and Laplace, who had carried the Newtonian theory to an extremely high pitch of detail, had enormous influence upon Ampère and his contemporaries. Newton's theory of gravitation was based upon the existence of a force of attraction between two particles which varied inversely as the square of their distance apart and was directed along the line joining the particles. This law had recently been extended to electrostatics and magnetism as the result of the experiments of Coulomb. It had been shown to apply both in respect of the magnitude of the force and its direction. As a result Poisson had been able to transform the status of the theory of electrostatics from that of an almost entirely qualitative science, by no means

---

†The sections in the commentary correspond to those into which the extracts from the papers, printed in this volume, have been grouped.

free from serious misconceptions, into an exact, quantitative affair with an elaborate mathematical structure. He was shortly to accomplish the same systematization for magnetism, and both these achievements represented the more or less direct application of the Newtonian theory as elaborated by Laplace and others. Inverse square laws were the order of the day. Since they corresponded with an emanation from the attracting body, they possessed some advantages from the point of view of ease of thinking and, moreover, had a compelling air of conviction about them.

The time was ripe for a similar advance in electrodynamics, but for some time events hung fire. This was because the requisite experimental procedure had not been hit upon, and until this had occurred progress was impossible. As Professor Hansteen wrote, a connection between magnetism and the electric current had long been suspected, but attempts to demonstrate it had all ended negatively. Knives in a case which had been struck by lightning had been found to have become magnetized, but the phenomenon, an isolated incident, had proved impossible to reproduce or investigate. The effect which was looked for was the orientation of an electric pile when suspended by a thread, but on open circuit. Oersted was still demonstrating negative experiments as late as 1819. Even when he tried the effects of closing the circuit and attempted to find some force between the circuit and a suspended magnet, he, unfortunately, held the wire at right angles to the magnet and so obtained no effect. When he did, in 1820, at last succeed, the publication of his results led to a veritable cascade of further work. It was the door for the opening of which the scientific world had been waiting. It is with this burst of development that the following papers are concerned.

We start with Oersted's paper which speaks for itself. It was published in Latin but was quickly translated into other languages. Its language is obscure. What was meant by the "electric conflict" which was taking place outside the wire conveying a current, is very difficult to imagine. It was something in the nature of a vortex motion, to which most bodies offered no resistance. Magnetic substances offered resistance to it and were carried along by it in consequence. It is, however, difficult to see any reason why one end of a magnet should be carried in one direction and the other in the opposite direction on this view.

B. BIOT AND SAVART

Arago, who had been travelling abroad, brought the news of Oersted's discovery to Paris. He described Oersted's experiments to a meeting of the French Académie des Sciences on 11 September 1820. From this Ampère concluded that it was possible that electric circuits might exert forces on one another, although this could not be deduced from Oersted's discovery. As Ampère pointed out, a magnet exerts forces upon a piece of soft iron but two pieces of soft iron are without effect upon each other. The only possibility of deciding this point was by means of an experiment, and this he performed immediately. Within a week of Arago's announcement of Oersted's results he had shown that two parallel wires carrying currents attracted each other if the currents they carried flowed in the same direction in both wires, and repelled each other if they were in opposite directions.

On 30 October 1820 Biot and Savart announced the results of experiments in which they determined the forces exerted on a small suspended magnet by a current in a long straight wire. It has to be remembered in reading Biot's account of the work of himself and Savart that the only source of electric current which was available to them was a battery of simple cells and that extreme difficulty must have been experienced in maintaining the current constant. There was no indicator to enable them to control this and, at the same time, they had to eliminate the effect of the earth's magnetic field.

The language in which they expressed their results was that appropriate to the then accepted theory of magnetism. This was the two-fluid theory, of austral and boreal magnetism respectively, already mentioned. A magnetic element comprised a molecule of austral and one of boreal magnetism, linked together.

On the basis of the results of Biot and Savart, Laplace deduced that the force exerted on a magnetic pole by an element of an electric current must follow a law of inverse squares. This result was not published by Laplace, but it is referred to in the writings of both Ampère and Biot himself.

Though Biot and Savart must have been extremely fine experimenters, they were less successful in the theoretical interpretation of their results, and they laid themselves open to the criticism of Ampère. Ampère attacked Biot particularly, partly because he neglected to acknowledge indebtedness which Ampère thought was due to him and to one or two other physicists (Savary, for example), and partly for committing several elementary errors. It is fairly clear from his writings that Ampère towards the end of the period had little use for Biot. Nevertheless, in spite of the fact that many of the criticisms which Ampère levelled against their conclusions were perfectly justified, there can be no doubt that the experimental work of Biot and Savart deserves inclusion in this collection of papers. It is, after all, their form of the law of action of current elements which has survived, rather than Ampère's own. Their experiments rank alongside those performed by Coulomb, in which forces are investigated by direct measurement. In conception they are not in the same class, however, as those designed by Ampère using null methods, upon which he based his theoretical studies.

Among the criticisms which Ampère levelled against the theoretical ideas of Biot and Savart, two may be mentioned. Speaking of a current element they had written:

"The nature of its action is the same as that of a magnetised needle which is placed on the contour of the wire in a definite direction and always constant relative to the direction of the voltaic current."

Ampère was quick to point out that a small magnet would exert a force which varied inversely as the cube of the distance and not the square, and that with the needle placed tangentially to the cross-section and at right angles to the axis of the wire, as it would have to be if it was to account for the direction of the forces which were found, a complete circle of such needles placed around the circumference of the cross-section, would exert no external force at all.

The second criticism which Ampère succeeded in bringing against Biot's interpretation of his results, referred to the second series of experiments in which he had investigated the effect of inclining the wire, and in which he had arrived at the perfectly correct law—that which is associated with his name and that of Savart. Biot had written:

"I have found that for the one as for the other [i.e. for the bent wire as for the long straight wire] the force is inversely proportional to the distance . . . but the absolute intensity was more feeble for the inclined wire than for the straight wire in the proportion of the angle of inclination to unity."

Ampère quotes M. Savary, who pointed out that to obtain the law, which as mentioned above is perfectly correct, it is necessary for the force to vary, not as the angle of inclination, but as the tangent of half that angle.

It is easy to see that this should be so. In their first experiments Biot and Savart had placed their small suspended magnet in the neighbourhood of a long straight vertical wire. In those which Ampère criticized they bent the wire at M on the level of the suspended magnet m into two arms ZM and MC (Fig. 1) still in the vertical plane through m but inclined at an angle α to the horizontal.

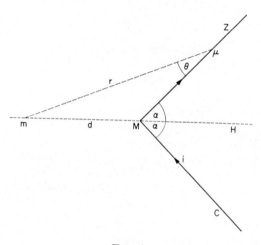

FIG. 1

Assuming the formula of Biot and Savart, which in slightly more more modern form is

$$B = K \frac{i.ds \sin \theta}{r^2},$$

the factor K being introduced to allow for any subsequent choice of units which may be desired, the force exerted on a unit pole placed at m by the current i in the two arms, would be

$$2K \int_{\theta=0}^{\theta=\alpha} i.ds \sin \theta / r^2 = 2K \int_0^\alpha \frac{id\theta}{r}$$

$$= 2K \int_0^\alpha \frac{i \sin \theta}{\mathrm{d} \sin \alpha} d\theta$$

$$= \frac{2Ki}{\mathrm{d} \sin \alpha} \left[-\cos \theta \right]_0^\alpha$$

$$= \frac{2Ki}{\mathrm{d} \sin \alpha} (1-\cos \alpha)$$

$$= \frac{2Ki}{\mathrm{d}} \tan \frac{\alpha}{2}$$

It is doubtful if the experiments of Biot and Savart would have been sufficiently accurate to decide whether the field was proportional to α or to $\tan \frac{1}{2}\alpha$.

The second paper reprinted in this volume is a mere announcement in the *Journal de Physique, de Chimie, d'Histoire Naturelle et des Arts*, of Biot and Savart's results, but it contains the passage leading to Ampère's first criticism. The passage is certainly muddled. It is not clear who actually wrote it but it is difficult to believe that it would have been printed without Biot or Savart being consulted.

Biot described their experiments in detail in his *Précis Elémentaire de Physique*. Ampère's second criticism was levelled against the account in the first edition. The passage which we give as the third reprint in the present volume is from the third edition of Biot's book in which the mistake had been corrected.

C. THE EARLY WORK OF AMPÈRE AND HIS THEORY OF MAGNETISM

We come now to Ampère's own contribution to the subject. We will consider it in five sections. The first deals with his preliminary experiments on the topic. The second is concerned with his general philosophy. In the third occurs his account of his main series of experiments upon which the theoretical discussion is based. The fourth comprises the deduction of the law of action of current elements, and the fifth his theory of magnetism.

From the first Ampère realized the importance of examining the question of the action of one electric circuit carrying a current upon another. It was he who first realized the possibility of such action existing and this was immediately after Oersted's results became known. He also realized the necessity for devising experiments to see if such action occurred. As soon as this was established qualitatively he again realized at once that if the action of a current upon another current, and the action of a magnet upon a current and of one magnet on another, were to be brought within the ambit of a single theory, it was to the action of current on current that attention must be turned. It was the only line of thought that promised to do anything beyond investigate the three separate cases as three distinct and unrelated phenomena. A theory which concentrated upon magnetic poles, or, as it was then, upon the two magnetic fluids composed of austral and boreal molecules, could do nothing towards a solution of this problem.

Ampère's line of reasoning appears to have been as follows. Oersted's experiments demonstrate that a magnet is orientated by an electric current. It is also orientated by the earth. Could it be that electric currents in the earth are responsible for the fact that a suspended magnet points in a north–south direction? Further, if this is so and the earth is a magnet by virtue of currents which circulate in it, could not circulating currents also explain the properties of ordinary magnets themselves? If this is so then electric currents should exert forces on each other. An experiment is required to see if this is actually the case.

The first selection from the early work of Ampère concerns these preliminary experiments. It is taken from two papers in the *Annales de Chimie et de Physique*. At the beginning Ampère has to attempt to distinguish between the effects of current and potential difference. In thinking of the latter he has electrostatic tension mainly in mind; he was only feeling his way towards an appreciation of the comparatively small potential differences which can exist in a conductor in which a current is flowing. Ohm's work still lay in the future (1826). Having invented a rudimentary galvanometer he describes his qualitative experiments on the attraction and repulsion of electric currents.

We omit his suggestion for an electric telegraph and end this selection from the first paper with his outline of his theory of magnetism. At this stage Ampère is obviously thinking of macroscopic currents rather than the molecular currents which he later proposed. The particles of the steel bar of a magnet acted like the elements of an electric pile and drove a current round the bar producing a solenoidal electric current. He had arrived at this idea from a similar postulate about the earth currents by means of which he explained terrestrial magnetism. In this case he imagined that the different rocks and minerals in the earth's crust acted like a pile generating currents in planes parallel to the equator. He even suggested that the heat of the earth might be caused by such currents.

The last paragraph, in which he relates the positions of the poles of a magnet to the direction of the circulating electric currents round the axis of the magnet, needs a figure. A swimmer *X*

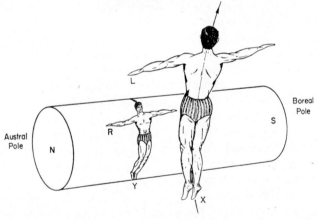

Fig. 2

in a wire, proceeding in the same direction as the electric current in it and facing a magnet which it deflects, sees the north-seeking pole—the pole of austral magnetic fluid—displaced towards his left hand. A corresponding swimmer *Y* in the circulating current of the magnet and facing the first, has the north pole on his right. If the first swimmer was also in the circulating current of a magnet he, too, would be looking outwards from his own magnet, so that in the position in which they are drawn in Fig. 2, they are similarly placed.

Ampère then proceeded to try to imitate the behaviour of magnets by means of electric currents flowing in helical coils of wire. Apart from the fact that he failed to get his helices to orientate themselves in the earth's field, he succeeded in establishing the analogy experimentally. He later succeeded in getting a current to orientate itself in the earth's field by using a single turn of platinum wire, lightly mounted, and it was this which was referred to as attracting to his laboratory many of the famous figures of physics.

In the extracts which we reproduce is included his reference to Biot's experiment with a hollow magnet. It is at least possible that the conclusion which he drew, namely that the currents in a magnet must be distributed throughout its volume, may have led him to his later idea of molecular currents.

These molecular currents he conceived as emanating from one end of a molecule and returning to the other through the surrounding space. Except that he does not envisage these return currents extending to infinity, his picture is almost precisely the same as that suggested by Heaviside for his "rational current element" towards the end of the century. A section across an Ampèrian molecule would correspond to Fig. 3. When unmagnetized the molecule would be symmetrical about its axis and the circulating currents

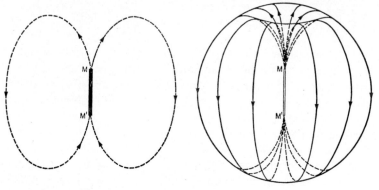

FIG. 3   FIG. 4

would return over a surface in the surrounding space, as in Fig. 4. The circulation of Fig. 4 would, in effect, be toroidal—like the circulation in a vortex ring.

As such it would produce no magnetic field at external points. It thus corresponds to an unmagnetized molecule, a picture which is magnetically completely equivalent to the modern one of balanced electronic orbits. A magnetized molecule in Ampère's theory, on the other hand, had the circulation displaced to one side, as in Fig. 5. By resolving these circulations into the sum of elementary circulations round the meshes of a reticulum, Ampère showed that the molecule would be equivalent to a smaller closed toroid, producing no external magnetic field, together with open-ended solenoids, going only partially round the molecule.

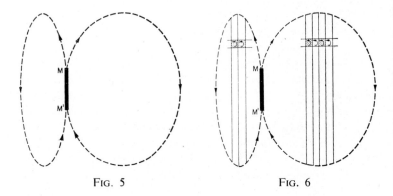

FIG. 5   FIG. 6

The latter would, of course, generate a magnetic field at external points.

In a magnetized specimen the axis of magnetization would be normal to the direction of the resultant circulation. The line joining what Ampère refers to as the ends of the molecule, which is named *MM'* in the figures, must therefore be aligned at right angles to the axis of magnetization. This was similar to the position of the molecules in Ampère's first picture of the circulating currents in a magnet, when he imagined the molecules as acting as elements in an electric pile. On this earlier view the molecules were, in fact, put together like Heaviside current elements and the molecular axes would lie tangentially to circles in planes parallel to the equatorial planes of the magnet, and whose centres lie on the magnetic axis of the magnet. The currents generating the magnetism would then flow along the circumferences of these circles.

D. AMPÈRE'S PHILOSOPHY OF SCIENCE

It is to the excellence of both his experimentation and the subsequent analysis of his results that Ampère's supremacy is due. In fact, his work lies at the root of all subsequent theories of electricity. It is impossible to read his paper without being profoundly impressed by his genius. It is perhaps a little unfair to compare him with Newton, whom he took as his model for a scientific investigator; he had Newton's existing success to build upon. Nevertheless, Maxwell has referred to him as the "Newton of Electricity" and to his law of action of current elements as the "Cardinal Principle of Electricity".

Ampère's paper was published in the *Mémoires de l'Académie des Sciences de Paris* (Vol. VI) for 1823. It is, indeed, one of the most fundamental publications in the history of the science of electricity. The volume, however, was not issued until 1827 and, when it was, the results of later work were incorporated in it—something which would not be tolerated today. The date of writing is mentioned in the text (p. 368). It is there given as 30 August 1826. The substance of it was read to the Académie at various stages, the last being in 1825. The paper extends to 213 pages and it is impossible to reprint it in its entirety. Some of it deals with the controversies of the time, the interest in which has long since subsided. Those parts can be omitted without loss. However, it still remains essential to exercise a considerable measure of selection if the matter is to be kept within bounds.

The principles upon which the selection from this paper were based have been to give sufficient to present Ampère's objectives and methods, to indicate his approach to both his experiments and the subsequent analysis, and to include those parts of his work which are fundamental to the subsequent development of the subject. From the latter point of view his law of action of current elements is of supreme importance. Upon it depends the possibility of representing the field set up by an electric current by means of continuous lines of force—in other words, the representation of the field by a non-divergent vector. Upon this, in turn, rest all subsequent field theories even though the theory which Ampère himself produced was one, like the theory of gravitation, of action at a distance. The other important contribution of Ampère, which he made in this paper, was to elaborate his theory of magnetism. Although this theory has in large measure remained, although not in detail, it has not the same fundamental importance as the law of action of current elements. The mathematics of this theory can be shortened a great deal. In such cases Ampère's line of argument is indicated, but the shorter form of the calculations is given.

The second selection dealing with Ampère's work (Part 2, Section 4D), which we shall now study, contains his discussion of the general principles which he followed. It is a statement of his philosophy of science, and merits some consideration. Though he set out to follow Newton in this, the line which he actually followed is, in some respects at least, rather in advance of those which Newton himself set out explicitly in the *Principia*.

At the beginning of the nineteenth century the influence of Newton was at its height. The theory of gravitation had been carried to an extraordinary pitch of exactness by Laplace and others, and the same general idea of an attraction following an inverse square law had been successfully applied to electrostatics, and a similar law had been demonstrated for magnetism by Coulomb. It was Ampère's ambition to do something of the same kind for current electricity, for the science of which he had coined the name of electrodynamics. Direct application of an inverse square law proved to be not immediately possible in this case, however, which turned out to be a good deal more complicated than either electrostatics or magnetism.

In setting about his investigations, Ampère modelled his methods consciously on those of Newton. The section reprinted here forms the introduction to the great Mémoire of 1823. In it Ampère sets out the principles he tries to follow. It was, of course, written *after* the conclusion of the work and represents Ampère's ideas on how his search had actually proceeded. The reader must decide for himself how much should be made of it. There is undoubtedly something in it but, at the same time, it is notoriously difficult to recapture a process of thought by introspection *a posteriori*, and in the writing of a paper it is very easy to rationalize the processes for the purposes of publication. How far this section represents Ampère's real train of thought and how far he was guided by other ideas and insights and preconceived notions, which he does not mention, is, of course, impossible to say. It is at least possible that he had the end in view, more or less clearly, from an early stage, and that he directed his experiments and theoretical investigations to the establishment of what he already felt must be the case.

The attitude of Newton, which Ampère sets out as his model, is one of severe empiricism, and it has since been the object of much criticism. Newton refused to admit that he had been guided by hypothesis at any point, and it was his famous dictum, "*Hypotheses non fingo*", which has brought on his head most of the subsequent criticism. The modern view is much more in accord with the saying attributed to Leonardo da Vinci: "Hypotheses are the general, experiments the soldiers." As Ampère writes as though he accepted Newton's position completely, what it involves deserves some examination.

To begin with it may be remarked that, whatever he may have said to the contrary, it is clear that Newton made very good use of hypothesis. His theory of gravitation is a theory of particles, or point masses, acting on each other at a distance, combined with laws of motion for such point masses. These are his initial postulates or hypotheses, from which he sets out, and on which he bases all his subsequent conclusions. These conclusions he then tested by comparison with experience, in so far as he already possessed it. That he did proceed in this manner is clearly shown by the holding up of his whole theory for a considerable period, because of a discrepancy between his estimate of the moon's motion and what he knew to be the observed facts, a discrepancy which was later removed by a redetermination of the earth's radius.†

Without question Newton made fundamental use of hypotheses and those mentioned in the previous paragraph are, by no means, the only ones which are implicit in his work. Ampère, working in the same spirit, will be found to have done the same thing.

---

†It has been suggested that Newton delayed publication until he had solved the problem of the action of a sphere at a distance and not because of uncertainty in the value of the earth's radius.

Nevertheless, it is worth trying to find out, in so far as it is possible, what Newton was attempting to say about his methods. He went out of his way to present a definition of what he meant by an hypothesis, and in this he took a rather different line from what would be followed today. The statement of his views occurs at the end of the *Principia*. Here he states:

"Hitherto we have explained the phenomena of the heavens and of our sea by the power of gravity. . . . But hitherto I have not been able to discover the cause of these properties of gravity from phenomena and I frame no hypotheses. For whatever is not deduced from the phenomena is to be called an hypothesis; and hypotheses, whether metaphysical or physical, whether of occult qualities or mechanical, have no place in experimental philosophy. In this philosophy particular propositions are inferred from the phenomena, and afterwards rendered general by induction." (*Principia*, 1713, p. 483.)

Nothing, of course, can be deduced solely from phenomena. Of themselves they possess no consequences—some general proposition about causation must be assumed in addition. What Newton seems to be saying is that his laws are to be looked upon as idealizations or generalizations from observations. His first law, for example, could be an idealization of the motion of bodies, such as balls rolling along grooves, or of ice sliding over a frozen pond. His second law might also be similarly assimilated to an idealization of the fact that when two equal forces appear to be acting (as, for example, by the extension of two similar springs) the acceleration produced is twice that found when only one of the forces acts. This process may be what he called "rendering general by induction", but it is by no means clear how he actually arrived at the second law in this way. For the third law he gives both an empirical and a deductive approach based upon the first two laws. Newton also accepted Kepler's laws as "phenomena". Given these and his laws of motion, the inverse square law of gravitation follows. He arrives at it by a simple assumption of circular motion for the planets. Then he says: "Now that we know the principles on which they depend, from these principles we deduce the motion of the heavens *a priori*." The logical pattern which he follows appears to be that of assuming the phenomena as a first approximation, deducing their consequences and then developing an accurate description *a priori*.

There is another sense in which Newton could claim to make no hypotheses. This lies in his studious avoidance of inventing a mechanism to account for gravity. He is content to work out the consequences of the fact that bodies behave *as if* they attracted each other by a force proportional to their masses and inversely proportional to the square of the distance between them. He was content to formulate the mathematical laws which follow and to leave their "explanation" to others or until later. This is a point which had a strong appeal for Ampère and he emphasizes it in his paper on the law of action of current elements.

In saying that in this sense Newton framed no hypothesis he could be thought of as feeling for what later developed into the positivist philosophy. What he seems to be trying to enunciate was not so very far removed from the principle contained in the verifiability definition of meaning or, indeed, the simple principle of Occam's razor, that entities should not be multiplied without necessity. He wished to rule out concepts, the introduction of which could not be justified by experience. Pressed further he would appear to be arguing against the use of models, which if not essential to scientific thinking, have at least been used with great success throughout the field of science.

To elucidate the position in more modern terms one might try to describe it on the following lines. A scientific theory consists of a set of statements and logical rules by which they can be manipulated, designed to provide a description of the pattern presented by the phenomena of nature, and by which it can be summarized and predictions about parts of the field so far unexplored, made. Strictly speaking it is independent of any model which may be employed as an aid to thinking about the phenomena. If we may take an example from later developments we might consider the dual wave-particle nature of light. The theory of light, strictly so-called, consists of the statements and rules by which the properties of light may be comprehended. In applying them it may be helpful to have the aid of a model and so, within a certain range it is possible to say that light behaves as if it were composed of streams of particles, while in another range its behaviour is as if it were composed of waves. In this case the model is ambiguous. The theory of light, however, is not ambiguous. Light is *sui generis*. The theory of light is the description of its properties, so far as they are known, by means of a system which can be operated by logical rules. In some cases, such as the kinetic theory of gases, for example, it is very difficult to disentangle the model from the theory, but even here it will be found that if the model is pushed to extremes it dissolves into unpicturable mathematical formulae. It was Newton's ambition to avoid such dilemmas by confining himself strictly to the description, in logical terms, of the phenomena. He did not wholly succeed, of course. The Newtonian particle has many of the ingredients of a model.

"Science is the attempt to make the chaotic diversity of our experience correspond to a logically uniform system of thought. In this system single experiences must be correlated with the theoretic structure in such a way that the resulting co-ordination is unique and convincing.

"The sense experiences are the subject matter. But the theory that shall interpret them is man-made. It is the result of an extremely laborious process of adaptation: hypothetical, never completely fixed, always subject to question and doubt." (A. Einstein, *Science*, 1940.)

It may be helpful to look a little more closely at the sort of logical system which both Newton and Ampère sought to find. The pattern which they followed is that of Euclid. Euclid did not discover his theorems, or at least not by any means all of them, by first convincing himself of the truth of his axioms and then working out their consequences, although that was the final form of his system. Many of the propositions which were embodied in his geometry had been known long before his time. The problem which he faced was much more nearly that of having been given the propositions, then determining what was the minimum number of postulates from which they could be deduced. In the theory of gravitation, Newton was already provided with a knowledge of a range of the phenomena, mainly through the medium of Kepler's laws. Ampère had to discover the laws as well as provide the theory, and thus do the work of Tycho Brahe, Kepler and Newton rolled into one. By themselves, although Kepler's laws provided a description of the phenomena, they were not incorporated into a logical system whereby deductions could be made. The problem which Newton so largely solved with his theory of universal gravitation, was the provision of such a logical system, based upon a limited number of postulates, by means of which the phenomena were made amenable to rational thought.

Before leaving this question we will glance briefly at the subsequent development of this position, since it helps in the understanding of what Ampère succeeded in doing in his theory of electrodynamics, and what has to be added to it if the theory is to remain adequate for modern purposes. The work of the mathematicians of the nineteenth and twentieth centuries led to a re-examination of this process of axiomatization. It was first brought into prominence through the introduction of geometries other than that of Euclid by the two Bolyais (father and son), Lobatchevski and Riemann, mainly in the first half of the nineteenth century. Euclid's axioms, which, until then, had been regarded as self-obvious propositions, were seen to be not the only set from which a consistent geometry could be deduced. The question which naturally arose as a result was, which of these possible geometries provided the true description of the world as we know it. The answer, as given by Poincaré, was that the geometry, pure and simple, should not be looked upon as a direct description of the world of experience at all. Rather it was a logical system in which the consequences of the initial assumptions were worked out; it had no necessary connection with reality which, with suitable adjustment in the statement of its laws, could be described in any geometry. The upshot of this view was the separation of the logical apparatus of the mathematics from the question of the physical experience. The logical apparatus became the province of the mathematics, while the problem of correlating the resulting logical system with experience, fell to physics. On this view the laws of mathematics became certain, in contrast with those of all the other sciences, which are always open to doubt and liable to revision. Mathematics, being subject only to the laws of thought thus achieved certainty, but at a price. The price was that mathematics became divorced from reality. To quote Einstein again:

"As far as the laws of mathematics refer to reality, they are not certain; and as far as they are certain, they do not refer to reality." (*Sidelights on Relativity*.)

Exactly the same distinction can be drawn in applied mathematics. In applied mathematics, strictly so-called, we would be dealing, according to this view, with bodies possessing postulated properties. The consequences of the possession of these properties are examined, but the question of whether or not bodies in the world of experience

possess such properties, is not discussed. A strictly positivist interpretation thus gives to applied mathematics the same certainty as to
pure, and at the same price—a divorce from reality. Naturally in
selecting the properties which they postulate for matter, mathematicians have an eye for what goes on in the world they know so
that their postulates are normally not far removed from what they
consider to be the world of experience. Nevertheless, interpreted in
this way, the task of deciding how the physical world is to be correlated with any of the possible mathematical worlds which can be
imagined, is assigned to physics.

The reader may well ask the question what does it matter whether
this process of correlation is called mathematics or physics. The two
branches were not separated in the mind of Ampère. There is no
doubt that he looked upon his mathematical investigations as
applying to the world of experience and not simply to a mathematical
universe of discourse. The distinction, all the same, has some
importance for us. Ampère's mathematical reasoning retains the
same elegance and exactness as it possessed when he developed it.
It is, as Maxwell put it, "perfect in form and unassailable in
accuracy". Maxwell went on to say, however, that "it is summed
up in a formula from which all the phenomena may be deduced".
This can only refer to Ampère's physics and not his mathematics,
and is a much more ambitious statement to make. There are advantages in keeping the mathematics and physics separate, the first
providing the logical machinery and the second determining the
accuracy with which the conclusions fit the world of experience.
Application can never be more certain than the correlation which
physics can provide. We may thus be able to improve the physics
while, at the same time, leaving the mathematics intact.

In the study of Ampère's papers we shall find that he bases his
theory on the results of certain experiments which he has performed.
The accuracy with which he was able to arrive at them was that
appropriate to the beginning of the nineteenth century, when
electrical experiments depended upon such instruments as the
electric pile. It cannot have been high. Whether we are prepared to
go all the way with positivism or not, Ampère's experiments can
hardly continue to be a basis with which we can be satisfied for the
continued acceptance of his conclusions today. What this basis can

be is a separate inquiry. The interest which lies in a study of the papers of the older physicists, such as Ampère, resides not so much in their furnishing us with the present basis of the science, as their portrayal of the genesis of the ideas which we still employ today. It is as examples of extremely skilful experimentation and ratiocination that they should be read and for which they are assembled in this book.

It is interesting to notice, in connection with the process of axiomatization of science, that, as will be seen in subsequent discussions, we have, in the theory which Ampère evolved for the action of current elements upon one another, on the one hand, and that first suggested by Biot and Savart for the same group of phenomena on the other, a very simple example of the possibility of correlating the same body of experience by means of two quite distinct mathematical systems. The two theories are based upon entirely different assumptions as to the forces which operate between elements of an electric circuit which are not, of course, open to direct experiment. Yet both are capable of accounting satisfactorily for the same facts. A good deal of confusion can arise through the existence of the two systems and particularly through a failure to realize that the method of application of the second is different from that of elementary Newtonian mechanics.

Newton was unusually sensitive to criticism. Hypotheses are often the easy target for adverse comment and part of his aversion to the employment of hypotheses, and particularly those assuming mechanisms which transcend the senses, may well stem from his desire to avoid controversy. The same can hardly be said of Ampère. Though he took Newton as his pattern, and was himself timid in the extreme, he did not show any strong desire to avoid speculation. In his investigations into the law of action of current elements he proceeded in a manner fairly closely akin to that of Newton, but half of his mémoire is directed towards his theory of magnetism, in which he postulated the existence of molecular currents. In the first he begins by establishing a number of facts by means of four basic experiments. These may be likened to those geometrical propositions which were known before Euclid's day, or to th laws discovered by Kepler, which provided Newton with the foundations for his theoretical considerations. He then seeks to unite them in a mathematical, axiomatic system corresponding to the Euclidean or Newtonian synthesis. The second part, dealing with the theory of magnetism, is not essentially different except that he there makes use of a mechanical model, that of molecular currents, to construct his theory. This model, together with its postulated properties, forms Ampère's system of axioms from which he starts. It is, of course, by no means self-obvious and is, indeed, a good deal more sophisticated than the basic assumptions he makes in connection with the interaction of current elements (though the resulting theory in the latter case is perhaps more sophisticated than his resulting theory of magnetism). To Newton, no doubt, the molecular currents would have been abhorrent and possessed of "an occult quality" out of place in a physical theory.

Running through the whole of nineteenth century physics was the feeling that it at last was unravelling "truth". The mechanical models represented real knowledge, and the real nature of matter was being deciphered. It is much to Ampère's credit that he does not push his hypotheses to such great lengths but, to quite a remarkable degree, he maintains an open mind. He is at pains to show that his theory of magnetism, based on solenoidal currents, gives essentially the same results as the then current theory of austral and boreal magnetic fluids. What he does legitimately claim for the theory is that it is the only one capable of comprehending the phenomena of magnetism, those of the action of currents on magnets and those of the mutual action of two or more currents, under one scheme. This remains true today and provides a very strong reason for adopting Ampère's line of thought in the development of the theory of electrodynamics by placing current electricity in the position of the basic science with magnetism arising from it as a subsidiary off-shoot. The theory of magnetism which is thus obtained is formally identical with that based upon the magnetic pole and developed along the lines of the theories of attractions in electrostatics or gravitation. To support the same case with the argument that magnetic poles "do not exist" is unsatisfactory. In the first place there is no means of knowing whether, in fact, elementary magnetic poles "exist" or not—on the large scale, at any level that admits the existence of magnets, there is no doubt of the existence of poles; and in the second, it is quite unimportant whether they exist or not. So long as it is known that electric circuits and magnets behave *as if* magnetic poles existed, it would be perfectly legitimate to employ the concept in magnetic theory. The real reason for proceeding otherwise is that advanced by Ampère. On the basis of elementary poles it is impossible to bring the phenomena into a single scheme. At least two quite distinct schemes would be required. However, even so, since the two theories of magnetic poles and molecular currents produce equivalent results, it would be perfectly legitimate in magnetism to pass from one point of view to the other as may be convenient. Any difference in convenience, however, can only inhere in the operator, through greater familiarity with one rather than the other, and not in the theory itself. Any result in the one theory can be reproduced as readily in the other, and for magnetism the two systems are indistinguishable. One, however, incorporates the magnetic action of electric currents and the other is incapable of doing so.

A further reason for making the basis of electrical theory rest in current electricity rather than in magnetism, and which some may consider the strongest reason of all, is that the necessary accuracy of measurement has not, so far, been attained in magnetism and, indeed, it seems most unlikely to be so. Our theoretical concepts ought to arise, so far as possible, from the most accurate measurements which have been performed. If they were to be based upon less accurate experiments, then their applicability to the more accurately investigated cases would remain uncertain. It is always possible, and indeed has usually been the case, that concepts derived originally on the basis of comparatively rough experiments have subsequently been refined as application has proceeded to other cases. The real basis of the concepts then shifts, in fact, from the initial observations in which they originated, and their justification is transferred to fresh experimental evidence.

The student of science is not in the position of having to follow the historical development of the subject willy-nilly. Many do, in fact, derive great interest and profit from following the genesis of the ideas they use, but whether or not this plan is adopted, all must arrive finally at an understanding of the present basis of the subject if they are to think rationally about it. It is no use employing concepts whose sole justification rests on experiments accurate to a per cent or so, in measurements requiring an accuracy of 1 in 100,000, without further consideration.

The law of action of current elements is a particular case in point. The measurement of electric current depends upon the current balance and is done to an accuracy of the order of 1 in 100,000. The theory of the current balance is a direct application of the law of action of current elements. It is necessary that this law, therefore, should be known to this order of accuracy. We shall return to this point later. It is mentioned here as a further argument for allowing the theory of current electricity to stand on its own feet. It again points to the advantages inherent in Ampère's approach.

E.   THE DESIGN OF AMPÈRE'S BASIC EXPERIMENTS

The problem which Ampère posed for solution was as follows. The mutual action between an electric current and a magnet, and between two electric currents having been discovered, what logical scheme can be devised to unify these phenomena quantitatively among themselves and also ally them with the much more extensive range already co-ordinated by Newtonian mechanics? In seeking a solution to this question Ampère argued as follows. Newtonian theory is a theory of action at a distance between particles, by-passing the question of the mechanism by which this action could be brought about. According to this theory the action of one particle on another is equal and opposite to that of the second on the first, and the line of action of both forces is along the line joining the particles. Thus any number of particles forming an isolated system cannot, by themselves, set the system in motion as a whole. The forces between the particles cannot be analysed any further; they are "elementary". In the case of the electric current the obvious analogue of the Newtonian particle is the current element. When, however, a small portion of an electric circuit is rendered mobile, it is found that the force which another current in a neighbouring circuit exerts upon it, is always directed at right angles to the isolated part, whatever the situation of the other circuit. Is it possible, asked Ampère, to find a law of force between the elements into which the two circuits may be divided such that, when the expression is integrated round one of the circuits, the force on the movable element in the other is always at right angles to that element, but which is such that the forces between the elements of the two circuits themselves, are always equal and opposite to each other and possess, for their line of action, the line joining the elements? This is the problem which he succeeded in solving and, although attention today is now directed to the magnetic fields generated by complete circuits rather than current elements, in doing so he produced a theory of the magnetic action of steady currents which remains intact to this day. He did, in fact, deduce the magnitude and direction of what we now call the magnetic field produced by a circuit—he called it the directrix rather than the magnetic field— and the law of its action on an element of current placed in it.

The form in which the law is employed today differs from that given by Ampère in the case of the current elements, though it is identical with it for the action of a complete circuit on an element of another circuit. The form which is in common use is that which goes by the name of the Biot–Savart expression, though Ampère deduced it also, in analytic form, for complete circuits. This is the well-known formula that the magnetic field produced by an element of a current $i.ds$ at a point distant r from it in a direction making an angle θ with the element, is $Ki.ds \sin \theta / r^2$ and the field lies in the direction perpendicular to both ds and r. The factor K is introduced to enable us to make any subsequent adjustment of units we wish. The force on a second element $i'.ds'$ is then given by the vector product of this field and $i'.ds'$.

It will be noticed that according to this law, the forces exerted by the current elements on each other are not necessarily equal, nor are they opposite, nor do they possess the same line of action. Consider two elements AB and CD at right angles and lying in the plane of the paper at a distance r apart, as in Fig. 7.

Fig. 7.

The magnetic field produced by AB at the second element DC is zero since the angle between r and AB is zero. AB, therefore exerts no force on CD according to the Biot–Savart formula. At AB, on the other hand, CD produces a field,

$$K\,\frac{i'.CD}{r^2}$$

The element AB, therefore, experiences a force equal to

$$K\,\frac{i.\,i'.AB.CD}{r^2}$$

in the direction of the arrow F. Newton's law of action and reaction is, thus, not obeyed between the elements, and it is to this that many elementary difficulties in using the Biot–Savart formula are due. When integrated round the complete circuit generating the magnetic field, however, the formulae of both Ampère and Biot and Savart give identical results. Since we never meet a magnetic field generated by isolated parts of a circuit in practice, it is impossible to distinguish experimentally between the two forms. In applying the Biot–Savart formula to parts of circuits, however, we have to remember that Newton's third law does not apply. If we make use of this formula we have to envisage that parts of circuits can exert forces upon themselves. To those thoroughly soaked in elementary Newtonian mechanics this may seem unreal, and it is necessary to be on guard about this point if mistakes are to be avoided. On the other hand, the use of Ampère's formula obviates this difficulty, since the mechanics he assumes agrees with that of Newton on this point, and makes the line joining the elements the line of action of the forces. The examples given in Part 1, Section F will make this clear.

In deciding to make this basic assumption about the line of action of the forces, Ampère was guided by his conception of "elementary" forces. He was well aware, for example, that two dipoles, such as two molecular magnets, act upon each other with forces which do not act along the line joining them. However, although the line of action of the forces in this case does not lie along the line joining the dipoles, the forces are, nevertheless, equal and opposite and they do possess the same line of action. But Ampère did not regard such forces as "truly elementary" since they could be decomposed into four separate actions arising between the respective poles, all of which do act along the lines joining the poles, in pairs. Was he, however, justified in assuming that the forces acting between current elements must necessarily be "elementary" in this sense? His assumption implies that current elements do not tend to rotate each other. Current elements, however, unlike Newtonian particles, possess direction as well as magnitude, and a turning effect ought not to be ruled out. Ampère assumed otherwise. He could justify his choice on the grounds that it was the simplest assumption to make and that, in the event, it proved adequate to account for all the phenomena. This could not have been foreseen at the beginning, however, and it is a point which he fails to discuss. It was taken up later by some of his critics.

It is true that Ampère's third experiment demonstrated that a complete circuit exerts a force on an element of a second circuit which is always at right angles to the element and that there is no couple tending to turn it in addition. This, however, does not answer the question, since it would be possible for *elements* of the two circuits to act on each other with couples so long as these couples neutralized each other when summed for a complete circuit. Such considerations are, therefore, insufficient to remove the difficulty of a turning effect produced by one element on another.

There is little that need to be said in introduction to Ampère's account of his experiments except, perhaps, to emphasize their remarkable ingenuity. Abandoning hope of the direct measurement of the forces acting between parts of circuits, all his experiments are based upon the possibility of securing situations of balance. It is the method of the "null" experiment. The previous experiments, such as those of Coulomb and of Biot and Savart, had all been direct. The introduction of the null method represented a considerable advance.

The four experiments which Ampère records in his paper are, obviously, not the only ones he has performed in this field. Rather they represent a tidying up of a large number of other experiments by means of which he was led to his theoretical ideas. They are just sufficient, and no more than sufficient, to establish the law of which he was in search. They show a conceptual elegance which could only be achieved *post hoc*. There is one very sad comment included at the end of the whole Mémoire which refers to the very crucial fourth experiment described in the extract given in Part 2, Section 4E.

"I ought to say [he wrote] in finishing this mémoire, that I have not yet had the time to construct the instruments shown in Figure 4 of the first Plate and in Figure 20 of the second." [This latter has little significance in the theory but that in Figure 4 is, in fact, his fourth experiment of the extract.] "The experiments for which they were designed have not yet been performed; but as these experiments have only the object of verifying results obtained otherwise and since they would be useful as a confirmation of those which have furnished these results, I have not felt that I ought to omit the description."

It is a pity that the state of scientific writing at the time had not sufficiently advanced for it to be considered necessary to record exactly what had actually been done, rather than some polished refinement. Ampère was rather too much concerned with the elegance of his account. He can certainly not be charged with dishonesty since, had he been dishonest, he would not have recorded this admission at all, but he cannot be absolved from a failure to record exactly what he had done.

It is by no means difficult to perform an experiment on the lines of that described in the Mémoire, which, without any great precautions, can easily exceed in accuracy anything which would have been likely to have resulted from Ampère's proposals. As a demonstration for a lecture the writer prepared two pairs of short solenoids. Each pair consisted of a small solenoid which hung inside the larger member of the pair, being supported from the arms of a chemical balance (see Fig. 8). The same current was led through both coils and the force on the suspended coil was measured for a series of positions of the fixed coil in the vertical direction. The maximum force experienced was determined from the results by a graphical method. The pair of coils was then replaced by the second pair, which was made to have twice the dimensions of the first and was wound with the same number of turns. When the same current was passed through this pair as had been used for the first, the maximum force experienced by the suspended coil for various positions of the fixed coil was the same as that obtained for the first pair, within an accuracy of 1 in 700. When both pairs were in the position corresponding to the maximum force they must have been in geometrically similar situations. Thus the force between any pair of circuits must be independent of the linear dimensions so long as they are similarly situated, as in Ampère's experiment with the circles moving in the horizontal plane, and his conclusions would follow. A photograph of the apparatus is reproduced in Plate III.

PLATE III. A demonstration experiment to show that the force between two current carrying circuits is independent of linear dimensions. A small coil is suspended from the arm of a balance inside a larger one and the maximum force experienced by the suspended coil determined by gradually raising the large coil. This pair of coils is then replaced by the smaller pair, shown near the balance case. The maximum force experienced in the two cases is the same if the currents are the same, although the dimensions of the smaller pair of coils are only one half those of the larger. (*By courtesy of School Service Reviews.*)

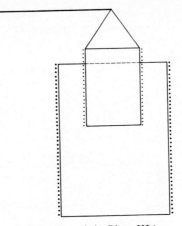

FIG. 8. (Photograph in Plate III.)

F. AMPÈRE'S THEORY OF THE ACTION OF CURRENT ELEMENTS

We now come to the theoretical edifice which Ampère constructed on the basis of his four experiments. As a result he arrived at his famous law of action of current elements. As has already been mentioned, the form in which he gave it is not that in which it is commonly used today. The latter is that known as the Biot–Savart form. How far was Ampère anticipated by these experimenters? The main facts, as far as they are ascertainable, appear to be as follows. We have already met the papers of Biot and Savart. There is no doubt that they experimented quite independently of Ampère and stated their result for the long straight wire and probably also for the inclined wires, without knowledge of the work of Ampère. Laplace's deduction of an inverse square law for current elements acting on magnetic poles, resulted from the first of these results and seems to have been arrived at by a method of dimensions. The error of calculation in the second series of experiments by Biot and Savart has been referred to; it is difficult to judge how serious this error was. It may have been simply a slip. On the other hand, they may have arrived at the correct result by means of an entirely faulty process of reasoning, on the basis of experimental results too rough to bring the error into prominence.

Biot and Savart throughout investigated the forces exerted by currents on magnets. Ampère was the first to experiment with the forces exerted by one current on another. His quantitative results seem to have been mostly obtained at a later date than those of Biot and Savart, but the precise order is impossible to unravel because of the long delays in the publication of results and the habit of incorporating new material into the Mémoires in the interval. It shows a fine disregard for priorities in publication which might, with advantage, become more widespread today, and it does not affect the subject itself, but it makes the task of attributing particular achievements to particular men extremely difficult.

Ampère was thinking along his final line of argument from a very early time. He first gave the formula

$$\frac{i.i'.ds.ds'}{r^2} \left\{ \sin\theta . \sin\theta' . \cos\omega + k \cos\theta . \cos\theta' \right\}$$

for the force between two current elements, making angles θ and θ' with the line joining them and the two planes containing this line and the two elements respectively making an angle ω with each other, in a paper communicated to the Académie on 4 December 1820. In this expression k was a constant he had been unable to determine. Perhaps the main achievement of the Mémoire of 1823 was the determination of the value of this constant as $-1/2$.

Many physicists were working at the same problem simultaneously by different methods. Among them Biot, Savart and Ampère were undoubtedly the chief, and it appears to the writer that the current convention of speaking of the Biot–Savart form of the law of action of current elements of Ampère provides as reasonable a recognition of the work of all three as can be attained.

We will postpone further consideration of the relation between the two forms of the law until after following the section of Ampère's paper, which sets out his theory. We will follow this for the moment only as far as his determination of the value of the constant k in his preliminary expression but first one or two examples in the use of Ampère's first formula.

Examples in the use of Ampére's formula

It will be seen from Part 2, Section 4F, that Ampère's complete formula finally emerges as,

$$F = \frac{i.i'.ds.ds'}{r^2} \left\{ \sin\theta . \sin\theta' . \cos\omega - \tfrac{1}{2} \cos\theta . \cos\theta' \right\}.$$

We need not go further into the alternative expressions which he evolves for the purposes of additional calculations. It will suffice if we give first two examples of the use of the formula to show that it does, in fact, give the same result as the Biot–Savart formula which everyone has been used to, in these cases, and discuss the so-called system of electrodynamic units which Ampère used.

Let us first take the case of the action of a long straight wire. Consider the force which a current i in the long straight wire AB (Fig. 9) exerts on a current element $i'ds'$ placed at Q in the direction cd, parallel to AB.

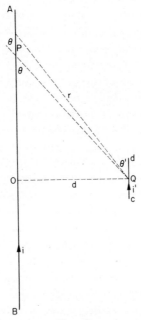

Fig. 9.

Consider the force which the element $i.ds$ at P exerts on that at Q, $i'.ds'$. We have $\omega = 0$. The force will lie along the line PQ and, since $\theta = \theta'$, its magnitude will be

$$\frac{i.i'.ds'}{r^2} \left\{ \sin^2 \theta - \frac{\cos^2 \theta}{2} \right\} ds.$$

Resolving in the direction OQ we find for the component in this direction,

$$\frac{i.i'.ds'}{r^2} \left\{ \sin^2 \theta - \frac{\cos^2 \theta}{2} \right\} \sin \theta . ds.$$

Since

$$r = \frac{d}{\sin \theta} \quad \text{and} \quad r.\frac{d\theta}{ds} = -\sin \theta,$$

we have for the total force which the long straight wire exerts upon the element $i'.ds'$, in the direction QO is,

$$\frac{i.i'.ds'}{d} \int_0^\pi (1 - \tfrac{3}{2} \cos^2 \theta) \sin \theta . d\theta$$

$$= \frac{i.i'ds'}{d} \left\{ \cos \theta - \frac{\cos^3 \theta}{2} \right\}_0^\pi$$

$$= \frac{i.i'}{d} ds'.$$

Had we used the form of the Biot–Savart formula expressed in electromagnetic units, as is usual, we would have obtained the well-known expression for the component of the force in this direction:

$$\frac{2i}{d} i'ds'.$$

The absence of the factor 2 in the result of the calculation when using Ampère's expression is caused by the fact that the latter used a different system of units, known as electrodynamic units. The unit of current in the electrodynamic system is thus smaller than that in the electromagnetic system by a factor of $\sqrt{2}$. The electrodynamic system has nothing to commend it and it has long fallen into disuse.

Proceeding with the calculation using Ampère's formula we find for the component of the force parallel to AB

$$\frac{i.i'ds'}{d} \int_0^\pi (\tfrac{3}{2} \sin^2 \theta - \tfrac{1}{2}) \cos \theta . d\theta$$

$$= \frac{i.i'ds'}{d} \left[\frac{\sin^3 \theta - \sin \theta}{2} \right]_0^\pi$$

$$= 0.$$

The force is, therefore, at right angles to cd, as we would have expected.

As a further example let us take the direction of the element $i'.ds'$ to be at right angles to AB but still lying in the plane ABQ, so that ω remains equal to zero (Fig. 10).

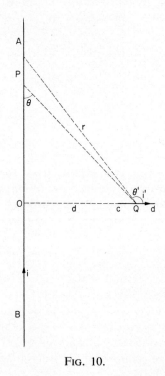

Fig. 10.

We find for the component of the force in the direction parallel to AB:

$$\frac{i.i'.ds'}{r^2} \int_{\pi}^{0} (\sin \theta \cos \theta + \tfrac{1}{2} \sin \theta \cos \theta) \cos \theta \, ds$$

$$= \frac{i.i'.ds'}{d} \int_{\pi}^{0} \tfrac{3}{2} \cos^2 \theta . \sin \theta . \, d\theta$$

$$= \frac{i.i'.ds'}{d} \left[\frac{\cos^3 \theta}{2} \right]_{0}^{\pi}$$

$$= \frac{i.i'.ds'}{d} .$$

This is the same force as was experienced by ds' when it lay in a direction parallel to AB, except for a change of 90° in its direction, which agrees with what we would expect from our previous experience using the Biot–Savart formula. It is easy to show that the component of the force at right angles to AB in this case becomes zero and thus that the force remains at right angles to the element.

When cd is placed in the direction at right angles to the plane AQB we have that $\omega = \tfrac{1}{2}\pi$ and $\theta' = \tfrac{1}{2}\pi$, so that

$$\sin \theta . \sin \theta' . \cos \omega - \cos \theta . \cos \theta' = 0$$

for all elements such as ds in AB. The total force on $i'ds'$ is thus also zero, which agrees again with the calculations from the Biot–Savart formula.

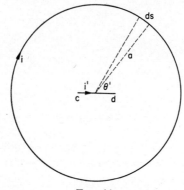

FIG. 11.

We will complete our examples of the application of Ampère's formula by calculating the force experienced by a current element $i'\mathrm{d}s'$ placed at the centre, and in the plane of, a circular current i, whose radius is a (Fig. 11).

Here we have: $\theta=\frac{1}{2}\pi$ for all elements $\mathrm{d}s$ of the circular current, $\omega=0$ and $a.\mathrm{d}\theta'=\mathrm{d}s$.

The force at right angles to cd will be,

$$\frac{i.i'.\mathrm{d}s'}{a^2} \int_0^{2\pi} \sin\theta'.\sin\theta'.\mathrm{d}s$$

$$=\frac{i.i'.\mathrm{d}s'}{a} \int_0^{2\pi} \left\{\frac{1-\cos 2\theta'}{2}\right\} \mathrm{d}\theta'$$

$$=\frac{i.i'.\mathrm{d}s'}{a} \left[\frac{\theta}{2}-\frac{\sin 2\theta}{4}\right]_0^{2\pi}$$

$$=\frac{\pi.i.i'.\mathrm{d}s'}{a}.$$

If we insert the factor 2 we convert the expression to one employing electromagnetic units, and obtain the usual familiar result.

If we resolve in the direction parallel to cd we obtain for the component of the force,

$$\frac{i.i'.\mathrm{d}s'}{a^2} \int_0^{2\pi} \sin\theta'.\cos\theta'.\mathrm{d}s$$

$$=\frac{i.i'.\mathrm{d}s'}{a} \int_0^{2\pi} \sin\theta'.\cos\theta'.\mathrm{d}\theta'$$

$$=\ 0.$$

Again, we find that the force experienced by cd is at right angles to it, as should be the case.

There are not many advantages to be obtained through the use of Ampère's expression in place of the Biot–Savart form. With the former, Newton's third law, concerning the equality of action and reaction, applies between current elements, and it is perhaps easier to avoid elementary mistakes on this score. The principal value in making a study of Ampère's theory is that it puts us on our guard when applying the Biot–Savart formula.

It is impossible to distinguish experimentally between these two forms for the law of action of current elements for the simple reason that it is never possible to produce a force on a conductor carrying a current by means of an isolated current element. The theory applies only to steady currents and these can only be obtained in complete circuits. For complete circuits the two expressions yield identical results. This we shall now proceed to show.

The Identity of the Formulae of Ampère and of Biot and Savart when Integrated for a Complete Circuit

Suppose that the element $i'.\mathrm{d}s'$, on which we wish to calculate the force exerted by another element $i.\mathrm{d}s$, is situated at P' (Fig. 12) and that it lies along the axis of z. Let the element $i.\mathrm{d}s$ be placed at the point P $(x, y, z.)$

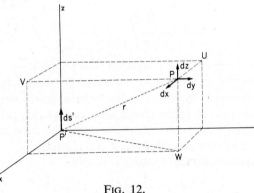

Fig. 12.

We shall use Ampère's expression in the form which involves the angle ε between the two elements themselves. Since

$$\cos \varepsilon = \sin \theta . \sin \theta' . \cos \omega + \cos \theta . \cos \theta',$$

Ampère's formula becomes

$$\frac{i.i'.\mathrm{d}s.\mathrm{d}s'}{r^2} \left\{ \cos \varepsilon - \tfrac{3}{2} . \cos \theta . \cos \theta' \right\}$$

or, if we wish to use other units in place of his electrodynamic ones, we may write his expression in the form

$$K \frac{i.i'.ds.ds'}{r^2} \left\{ 2 \cos \varepsilon - 3 \cos \theta . \cos \theta' \right\}$$

For Ampère's electrodynamic units $K=\frac{1}{2}$; for electromagnetic units $K=1$; for m.k.s. units $K=10^{-7}=\dfrac{\mu_0}{4\pi}$.

Referring to Fig. 12 we see that

$$\cos \varepsilon = \frac{dz}{ds}, \quad \cos \theta = \frac{dr}{ds}, \quad \text{and} \quad \cos \theta' = \frac{z}{r}$$

The component of the force given by Ampère's expression in the z-direction is, therefore,

$$F_z = K \frac{i.i'.ds.ds'.z}{r^3} \left\{ 2 \cdot \frac{dz}{ds} - 3 \cdot \frac{z}{r} \cdot \frac{dr}{ds} \right\}$$

$$= K \, i.i'.ds' . \frac{d}{ds} \left\{ \frac{z^2}{r^3} \right\} . ds.$$

When this expression is integrated round the circuit of which ds forms a part, this vanishes. This proves for the general case what we have already seen in the special cases we have studied so far, that the force exerted by a complete circuit on a current element is at right angles to the element. It was one of the major objectives of Ampère's investigation, of course, to find a law of this kind.

The component parallel to the axis of x is,

$$F_x = K \frac{i.i'.ds'.x}{r^3} \left\{ 2 \cdot \frac{dz}{ds} - 3 \cdot \frac{z}{r} \cdot \frac{dr}{ds} \right\} ds.$$

$$= Ki.i'.ds.ds' \frac{d}{ds} \left\{ \frac{x \, z}{r^3} \right\} + \frac{Ki.i'.ds.ds'}{r^3} \left\{ x \frac{dz}{ds} - z \frac{dx}{ds} \right\}$$

On integration round the circuit s, the first term vanishes. The remainder is the Biot–Savart expression for the x-component.

To see that this is so we note from the figure that a force in the x-direction is produced, according to the Biot–Savart formula, by the components dz and dx.

The element $i.dz$ produces a force

$$K \frac{i.i'.ds'.dz.P'W}{r^3}$$

and its component in the direction of x is,

$$K \frac{i.i'ds'.dz.x}{r^3}$$

The element $i.dx$ produces a force in the direction xP'

$$K \frac{i.i'.ds'.dx}{r^3} P'U. \frac{z}{P'U} .$$

The total component in the direction of x is, therefore,

$$K \frac{i.i'.ds'}{r^3} (x.dz - z.dx) = K \frac{i.i'.ds}{r^3} \left\{ x \frac{dz}{ds} - z \frac{dx}{ds} \right\} ds,$$

which is identical with the non-vanishing part of the expression deduced from Ampère's formula. A similar argument applies to the y-component.*

It is in this analytical form that Ampère himself deduces the Biot–Savart law, the correctness of which for complete circuits he unhesitatingly acknowledges, though the theory of these experimenters received, as we have seen, short shrift at his hands.

Since experience is unable to distinguish between the two forms of Ampère's law, it is perfectly legitimate to employ one or the other, according to convenience. When using the Biot–Savart form we can avoid an open infringement of Newton's third law if we refrain from statements about current elements and restrict ourselves to the integrated form, applied to complete circuits. For purposes of calculation we may say that closed circuits carrying electric currents

---

*I am indebted to Mr. R. C. Lyness for this demonstration of the identity of the formula of Ampère and of Biot and Savart.

behave as if each element of the circuit generates a magnetic field

$$K \frac{i.\mathbf{ds} \wedge \mathbf{r}}{r^3},$$

where $\mathbf{ds} \wedge \mathbf{r}$ represents the vector product of $\mathbf{ds}$ and $\mathbf{r}$, and the force experienced by any current element $i'.\mathbf{ds}'$ placed in this field is given by the vector product of $i'.\mathbf{ds}'$ and $K \dfrac{i.\mathbf{ds} \wedge \mathbf{r}}{r^3}$.

A Simple Deduction of the Biot–Savart Formula from the Experimental Results of Ampère

It may be of interest here, to give an account of a very simple method of deducing the Biot–Savart formula for a complete circuit on the basis of Ampère's original postulates and his four experiments. Any electric circuit may be divided into *circuit elements* by wires forming a reticulum bounded by the circuit—a device that was employed by Ampère himself. Matters are very considerably simplified if we investigate the force exerted by such a *circuit* element on an element of current, instead of the force between two *current* elements. We shall, in fact, need only to examine the case when the current element passes through a point in the plane of the circuit element.

Fig. 13.

The force exerted by a circuit element must be proportional to its area α since it can be divided into equal sub-elements all sensibly equidistant and similarly situated, as viewed from external points, so that they must all give rise to equal forces. It follows from Ampère's fourth experiment that the force on an element $i'\,ds'$, situated at a distance r, must vary as

$$K \frac{i.i'.\alpha.ds'}{r^3}$$

This follows by the method of dimensions which he himself employed.

Consider first a current element $i'\mathrm{d}s'$ which lies wholly in the plane of α. Assuming, with Ampère, that Newton's third law is obeyed, the force experienced by the current element must also lie in the same plane. We first prove, on the basis of Ampère's third experiment, that the force on the current element $i'.\mathrm{d}s'$ is independent of the direction of the element in the plane of α. The current element $i'.\mathrm{d}s'$ may be decomposed into a pair of elements $i'.\mathrm{d}p$ and

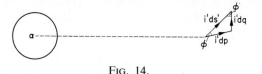

Fig. 14.

$i'.\mathrm{d}q$ chosen at random, making angles of φ and φ' with $\mathrm{d}s'$ respectively (Fig. 14). Suppose that the forces experienced by these new elements were $K_p.\mathrm{d}p$ and $K_q.\mathrm{d}q$ respectively. Resolving parallel to $\mathrm{d}s'$ and equating the resultant to zero by virtue of Ampère's third experiment, which showed that there was no component parallel to an element, we have

$$K_p.\mathrm{d}p \sin \varphi = K_q.\mathrm{d}q \sin \varphi'.$$

But

$$\frac{\mathrm{d}p}{\sin \varphi'} = \frac{\mathrm{d}q}{\sin \varphi},$$

giving

$$K_p = K_q.$$

If we apply a similar argument to an element inclined to the plane of α, we can show that the force on any element at right angles to this plane is zero. Consider the element $i'.\mathrm{d}s'$ in Fig. 15. Resolve it

Fig. 15.

into components $i'.\mathrm{d}z$ perpendicular to the plane of α and $i'.\mathrm{d}x$ parallel to that plane. The force experienced by $\mathrm{d}x$ will, as before, lie in the plane of α and therefore it must be in the direction of y. It has no component parallel to $\mathrm{d}s'$. The force experienced by $\mathrm{d}z$

must, therefore, also possess no component parallel to ds'. It must, therefore, also lie along the direction of y or be zero. By choosing an element ds' lying in the yz plane, it follows that the force experienced by dz must lie in the direction of x or be zero, from which it follows that the force must be zero.

The force on any current element $i'.$ds', passing through a point in the plane of α is thus proportional to $(i.\alpha)/r^3$ and to the resolved part of $i'.$ds' in the plane of α, and it is at right angles to the resolved part of ds' in that plane. If n is the unit vector in the direction at right angles to the plane of α, we have that the force experienced by i'ds' is

$$K\frac{i.i'.\alpha}{r^3}\, n{\wedge}\textbf{ds},$$

where K is our usual constant, depending upon the units chosen for the measurement of current.

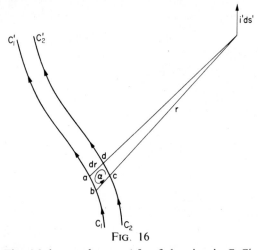

FIG. 16

Let ab, Fig. 16, be an element $i.$ds of the circuit $C_1C'_1$, for which it is desired to calculate the force exerted on the current element $i'.$ds' at P. Introduce the small circuit element $abcd$, round which a current i flows in the direction of the arrow. It will introduce a force on $i'.$ds' of

$$K\frac{i.i'.\alpha}{r^3}\, n{\wedge}\textbf{ds}'$$

If we introduce similar circuit elements all round the circuit C, keeping the ratio dr/r the same for all, the effect will be simply to shift the circuit from the position $C_1C'_1$ to $C_2C'_2$, the radial components of the current along bc and da being cancelled by the current in the neighbouring elementary circuits. The new circuit in the position $C_2C'_2$ will be similar to $C_1C'_1$ and similarly placed with regard to P. Hence Ampère's fourth experiment shows that if the circuit C', of which $i'.\mathbf{ds}'$ is a part, is also similarly diminished, the force which it experiences will remain unaltered. (Strictly speaking, Ampère's fourth experiment demonstrated this fact for a complete circuit C', but a reduction in scale cannot alter the direction of any of the forces but only their magnitudes in a constant ratio, and Ampère's result must apply to the elements of which the second circuit is composed). If $i'\mathbf{ds}'$ is not changed, therefore, the force on it must increase by a fraction dr/r of its original value, and this must be true for all elements, since the angles remain unaltered.

Hence, if we add up vectorially all the forces

$$K \frac{i.i'.\alpha}{r^3} \, \mathbf{n} \wedge \mathbf{ds}'$$

exerted by the circuit elements, such as $abcd$, the result will be $\mathbf{F}.dr/r$, where $\mathbf{F}$ is the force originally experienced by $i.\mathbf{ds}'$.

That is, therefore,

$$\frac{\mathbf{F}.dr}{r} = \sum \left[\frac{Ki.i'.\alpha.\mathbf{n}}{r^3} \right] \wedge \mathbf{ds}'.$$

Writing $\alpha = dr.ds \sin\theta$, and cancelling by the factor dr/r, which remains constant throughout, we obtain

$$\mathbf{F} = \sum \left[\frac{Ki.i'.ds \sin\theta.\,\mathbf{n}}{r^2} \right] \wedge \mathbf{ds}'$$

which is the Biot–Savart formula. The vector

$$\sum \left[\frac{Ki.ds \sin\theta.\,\mathbf{n}}{r^2} \right]$$

is that now called the magnetic induction.

The Equivalent Magnetic Shell

Before we leave the consideration of the various forms in which it is possible to express Ampère's law of action of current elements we will examine its relation to the expression for the magnetic potential at a point in terms of the solid angle subtended by a circuit at that point. Among the achievements which Ampère recorded in his famous Mémoire was the development of the theory of the equivalent magnetic shell. This he did by means of the lengthy mathematical analysis for which reference must be made to the paper itself. The result, however, is very simple to obtain from the Biot–Savart form of the expression for Ampère's law. We shall need it for a consideration of Ampère's theory of magnetism and it will be convenient to consider the deduction at this point. The result follows immediately from simple geometrical considerations, the only difficulty in seeing this is in interpreting the figure. What we have to show is that the magnetic field generated by a complete circuit, as given by the Biot–Savart expression, is equivalent to the gradient of a quantity which is proportional to the solid angle subtended by the circuit at the point being considered. That is,

$$\sum K \frac{i}{r^3} \, \mathbf{d}s \wedge \mathbf{r} = Ki \, \text{grad} \, \omega,$$

where ω is the solid angle subtended by the circuit.

Let us calculate the component of the field at a point P in some direction ρ (Fig. 17).

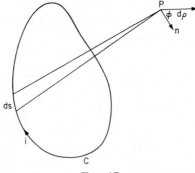

FIG. 17.

Let us suppose P to be displaced a distance dφ in this direction, and let us calculate the product of the component of the field in this direction and dφ. Instead of displacing the point P in the direction of φ we may, if we wish, displace the circuit C parallel to itself by the same distance dφ in the opposite direction (Fig. 18). Figure 19 shows an enlargement of the displacement of the element $i.\text{d}s$.

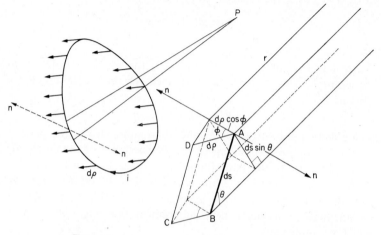

Fig. 18.   Fig. 19.

From the Biot–Savart formula the field at P will be,

$$\boldsymbol{B} = \sum K \frac{i.\text{d}s \sin \theta}{r^2} \cdot \boldsymbol{n}$$

Here **n** is the unit vector in the direction of the normal to the plane containing r and ds.

The component of $\boldsymbol{B}$ in the direction of φ will be

$$B_\rho = B \cos \varphi = \sum K \frac{i.\text{d}s \sin \theta}{r^2} \cdot \cos \varphi . \boldsymbol{\rho},$$

ρ being the unit vector in the direction of **ρ** and φ the angle between **ρ** and the normal to the plane defined by ds and r.

$$B_\rho \text{d}\boldsymbol{\rho} = \sum K \frac{i.\text{d}s \sin \theta}{r^2} \cdot \cos \varphi . \text{d}\boldsymbol{\rho}.$$

During the displacement $d\rho$, the current element $i.ds$ moves from the position AB (Fig. 19) to CD. Now it will be seen from the figure that the quantity,

$$ds.d\rho \sin\theta.\cos\varphi$$

is an area equal to the projection of the area $ABCD$ on to the plane at right angles to the radius vector r from P.

Thus
$$\frac{ds.d\rho \sin\theta.\cos\varphi}{r^2}$$

is equal to the solid angle subtended by the small area $ABCD$ at the point P. Thus when this is summed for the whole circuit we obtain,

$$B_\rho d\rho = K\,i.d\omega,$$

where $d\omega$ is the change in the solid angle subtended at the point P by the circuit, corresponding to the displacement of the circuit $d\rho$ or, what comes to the same thing, the displacement of the point P by the same amount in the opposite direction. It follows from this that the magnetic induction, generated at a point by an electric circuit, may be obtained from the gradient of a potential function which is simply equal to

$$K.i.\omega.$$

Thus $$\mathbf{B} = K.i \text{ grad } \omega,$$

the actual sign depending upon the convention employed in determining the sign to be attached to the solid angle ω.

The reader will be well aware that this potential function, $K.i.\omega$, is applicable only in cases where the currents are confined to definite wire circuits and the field of induction is being investigated outside the conductors carrying the currents. This potential is not single valued. The integral of $\mathbf{B}$ taken round a closed circuit is only zero if the path does not interlace current. If it does, the potential will not return to its original value since the solid angle will continually grow or diminish according to the direction of the path, all the way round. In fact, of course, if $\mathbf{B}$ is the gradient of a scalar potential it follows that curl $\mathbf{B}$ must be zero since the curl of any gradient vanishes identically, and as we see in the following paragraph, curl $\mathbf{B}$ is only zero in those regions where there is no current flowing.

Nevertheless, there are many problems, for example in connection with electrical measuring instruments and current balances, where this condition is satisfied and the derivation of the magnetic induction from a potential can be employed.

There are two important differences between magnetic and electrostatic potentials. In the first place, as we have just seen, the magnetic potential is not single-valued whereas the electrostatic potential is. In the second-place electrostatic potential is a scalar quantity whereas the magnetic potential is a pseudo-scalar. The sign to be attached to the magnetic potential depends upon the direction of circulation of the current giving rise to it—that is upon whether it is right- or left-handed. If the coordinate system is reflected in a plane mirror, the sign of the magnetic potential changes. The electrostatic potential, on the other hand, is not changed by such a transformation. The difference corresponds to the difference between polar and axial vectors. The ordinary vector, of which the paradigm case is the displacement of a point in space, is called a polar vector and is unchanged by all orthogonal transformations of the coordinates—that is by translations, rotations and reflections. Axial vectors, on the other hand, which contain the element of right-handedness or left-handedness, transform like ordinary vectors under translation and rotation of the coordinate system, but in the case of reflection the relative sign between axial and polar vectors is reversed. Examples of axial vectors are torque, angular momentum and magnetic induction. Examples of polar vectors are velocity, acceleration, force, and electrostatic induction.

The First Law of Circulation—the "Work Rule"

It was from the foregoing result that the well-known "work rule" arose. If the vector B is integrated around a path which encloses a current i flowing in a circuit, the solid angle which gives the potential from which B may be calculated will change continuously throughout and by the time the original point of departure is regained the total change will have amounted to 4π. From this it follows that

$$\int B.\mathrm{d}s = 4\pi K.i.$$

If the path of integration does not interlace current, $\int B.\mathrm{d}s = 0$. We postpone consideration of what modifications are required in the presence of magnetic material until after Ampère's theory of magnetism. This fundamental result is thus a direct consequence of Ampère's law of action of current elements.

If we have more than one such current the results are additive. In a region where current flows, therefore, since any closed path may enclose current, if we integrate B round each of the elementary rectangles, $dx.dy$, $dy.dz$ and $dz.dx$ in turn, we must equate the result to the current flowing through the elementary area, multiplied by $4\pi K$. If the current density is represented by the vector j, we obtain in the usual way,

$$\text{curl } B = 4\pi Kj.$$

The vector B can then no longer be the gradient of a scalar potential.

The Vector Potential

When the electric currents do not flow in paths which are restricted to known wire or other conducting circuits, but flow throughout space, we can return to Ampère's law in differential form and use it to establish a vector potential. Even in such cases steady electrical currents will flow in closed paths and we will be able to avoid the difficulties inherent in the differential form of Biot and Savart, if we always integrate over the whole of space, in which current flows.

Let us consider an element of current $i'.ds'$ placed at the origin of coordinates and pointing along the axis of z, as in Fig. 12. The magnetic induction at any point $P(x, y, z)$ is, in magnitude,

$$K \frac{i'.ds' \sin \theta}{r^2} = Ki'.ds'. \frac{PW}{r^3}$$

The component in the direction of x will be

$$B_x = K i'.ds'. \frac{PW}{r^3} \cdot \frac{y}{PW}$$

$$= K i'.ds'. \frac{y}{r^3},$$

and that in the direction of y will be

$$B_y = K\,i'.ds'.\frac{x}{r^3}.$$

Also, $$B_z = 0$$

Since $$r = \sqrt{(x^2+y^2+z^2)}$$

$$B_x = K\,i'.\frac{d}{dy}\left\{\frac{ds'}{r}\right\}$$

$$B_y = -\,Ki'.\frac{d}{dx}\left\{\frac{ds'}{r}\right\}$$

$$B_z = 0.$$

Since $i'.\mathbf{ds'}$ is a constant vector possessing no components in the direction of x or y we have for the x-component of curl $(i'.\mathbf{ds'}/r)$

$$\frac{d}{dy}\left\{\frac{i'.\mathbf{ds'}}{r}\right\},$$

and for the y-component,

$$-\frac{d}{dy}\left\{\frac{i'.\mathbf{ds'}}{r}\right\},$$

and the z-component is zero.

Thus if we write

$$A = K\frac{i'.\mathbf{ds'}}{r}$$

$$B = \text{curl } A.$$

When current is distributed throughout space we must replace the linear element of current $i'.\mathbf{ds'}$ in terms of the current density j. If, at the origin, this flows in the direction of the axis of z, $i'\mathbf{ds'}$ must be replaced by $j.dx.dy.dz$, that is by $j.dv$, where dv is an element of volume.

Thus the contribution towards A made by this element of volume will be

$$K \frac{j.\mathrm{d}v}{r} \, ,$$

and

$$A = \int K \frac{j.\mathrm{d}v}{r} \, .$$

Comparing this with the expression for the electrostatic potential V generated by a distribution of electric charge of density ρ,

$$V = \int \frac{\rho.\mathrm{d}v}{r} \, ,$$

we see that the vector A will satisfy Poisson's equation,

$$\nabla^2 A = -4\pi K j.$$

Also, since

$$\mathrm{curl}\, B = \mathrm{curl}\,\mathrm{curl}\, A$$
$$= \mathrm{grad}\,\mathrm{div}\, A - \nabla^2 A$$
$$= 4\pi K j.$$

Thus $\mathrm{grad}\,\mathrm{div}\, A = 0,$

which shows that div A is constant. From this it does not follow that div A need be zero. However, we can always add to A any vector whose curl is zero, the divergence of which need not vanish. In other words, it is always possible to choose A so that div $A = 0$. When this is done A is known as the vector potential.†

---

†This choice of A making div $A = 0$ is, of course, arbitrary. It is, in fact, no longer made when non-stationary phenomena are brought into consideration. In order to obtain relativistic covariance it is found that a different choice has to be made depending upon what are known as "gauge" transformations. Magnetic and electric fields are no longer independent but become different aspects of the same thing. In this book, however, attention in confined to steady currents where these questions do not arise, and the choice of A so that div $A = 0$ is the one conventionally made.

The Representatation of the Magnetic Field of an Electric Current by means of Lines of Force. The Magnetic Induction a Non-divergent Vector.

Before it is legitimate to represent the field of induction set up by an electric circuit in its neighbourhood, by means of lines of induction, it is necessary to show that magnetic induction generated in this way possesses the same properties as that produced by magnetic poles. In the case of magnetism these properties can be developed from the inverse square law and Gauss's theorem, which depends upon it. In the case of fields generated by electric currents, it is Ampère's law of current elements which has to take this place. In other words, it is necessary to show, on the basis of Ampère's law, that the magnetic induction set up by an electric current can be represented by a non-divergent vector. Armed with the solid-angle theorem given above, it is not difficult to verify directly, the non-divergence of the vector B at points external to the circuit.

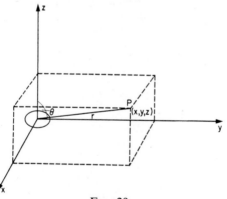

Fig. 20.

To do this it is sufficient to consider a small element of area α, of the reticulum into which any electric circuit may be divided. Take the origin at the centre of such a small area and let the direction of the axis of z be that of the normal to the area α.

The solid angle subtended by the area α at a point $P\,(x,\,y,\,z)$ is,

$$\omega = \frac{\alpha \cos \theta}{r^2} = \frac{\alpha . z}{r^3}\,.$$

We have,

$$r = \sqrt{(x^2 + y^2 + z^2)}$$

and

$$\frac{dr}{dx} = \frac{x}{r} \quad \frac{dr}{dy} = \frac{y}{r} \quad \frac{dr}{dz} = \frac{z}{r}$$

so that

$$\frac{d\omega}{dx} = \frac{-3\alpha z}{r^4} \cdot \frac{dr}{dx} = \frac{-3\alpha zx}{r^5}$$

$$\frac{d\omega}{dy} = \frac{-3\alpha zy}{r^5}$$

$$\frac{d\omega}{dz} = \frac{-3\alpha z^2}{r^5} + \frac{\alpha}{r^3}$$

Hence the three components of the magnetic induction at P are,

$$B_x = \frac{-3\alpha zx}{r^5} . K.i$$

$$B_y = \frac{-3\alpha zy}{r^5} . K.i$$

$$B_z = \left(\frac{-3\alpha z^2}{r^5} + \frac{\alpha}{r^3} \right) K.i.$$

Therefore,

$$\frac{1}{K.i} \frac{dB_x}{dx} = \frac{15\alpha zx^2}{r^7} - \frac{3\alpha z}{r^5}$$

$$\frac{1}{K.i} \frac{dB_y}{dy} = \frac{15\alpha zy^2}{r^7} - \frac{3\alpha z}{r^5}$$

$$\frac{1}{K.i} \frac{dB_z}{dz} = \frac{15\alpha z^3}{r^7} - \frac{6\alpha z}{r^5} - \frac{3\alpha z}{r^5} .$$

Whence we have by addition,

$$\text{div } \boldsymbol{B} = \frac{dB_x}{dx} + \frac{dB_y}{dy} + \frac{dB_z}{dz} = 0.$$

The magnetic induction generated by an element of an electric circuit will, therefore, be non-divergent, a property which will not depend upon the particular position which we have chosen to place it in for the purposes of the above calculation. The result will apply to all elements of a circuit and consequently to the circuit as a whole.

This result is of the greatest importance. It shows that the representation of the field of induction produced by an electric current by continuous lines of force, is not, in any way, dependent upon a law of inverse squares for magnetism. It was the development of the theory of magnetic fields based on the law of inverse squares for magnetic poles, to which must be attributed the fundamental place accorded to magnetism in electrical theory. Electric currents were replaced by their equivalent distribution of magnetism. It is, of course, still perfectly legitimate to do this, but when it is done, however, it is to the investigations of Ampère to which we must turn to justify the process. Nevertheless, this seems a very round-about way to achieve something which follows without employing a theory of magnetic poles at all. It involves invoking experimental evidence for an inverse square law for magnetic poles which is strictly not relevant at all, and with which it is possible easily to dispense. Moreover, the inverse square law for magnetic poles cannot be established to anything like the accuracy which is necessary for measurements in current electricity and so it cannot very well be taken as the basis for the theory of such measurements. Ampère's suggestion for the unification of the whole theoretical structure on the basis of current electricity is, of course, the solution which was ultimately adopted, though for curious reasons of historical inertia the result is usually only attained through a preliminary and unnecessary excursion into the theory of magnetism. In so far as this basis in magnetism is implied to be necessary, it is misleading.

Once we know that the magnetic induction B generated by electric currents is non-divergent, it becomes possible to apply Gauss's theorem. Lines of induction cut the surfaces defined by a constant value of ω, at right angles. Applying Gauss's theorem to a portion of a small tube of induction, generated by lines of force and bounded by two surfaces $\omega=$constant, we find that $B_1\sigma_1,=B_2\sigma_2$, where B_1 and B_2 are the values of the magnetic induction at the two ends and σ_1 and σ_2 are the areas of cross-section of the tube at these points. It is thus possible to represent a field of induction produced by any system of electric currents, in magnitude as well as direction, in the usual way, by drawing the number of lines per unit area of cross-section of a tube proportional to the magnitude.

FIG. 21.

Subsequent development of electromagnetic theory was based upon theories of fields. The advantages of a field theory are not apparent when considerations are restricted to steady currents. It is perhaps worth noting, however, that fields are never experienced directly in themselves. The sole way in which they are made manifest is by actions which are experienced by bodies placed in them. This is as true of light and electromagnetic radiation generally as of the magnetic fields generated by stationary currents. Fields, when looked upon as entities in themselves, are mental concepts useful for the interpretation of the behaviour of bodies more directly accessible to the senses, like many other hypothetical entities employed in physics. It is not very profitable to debate whether or not they "exist".

Units and Dimensions

We will conclude this section by a further reference to the question of units, starting from Ampère's law of action in its Biot–Savart form. The force exerted by an element of current $i.\mathbf{ds}$ upon another, $i'.\mathbf{ds}'$ at a distance r is looked upon as being given by

$$ \mathbf{F} = K \left(\frac{i.\mathbf{ds} \wedge \mathbf{r}}{r^3} \right) \wedge i'.\mathbf{ds}'. $$

For reasons depending upon the impossibility of isolating the element ds producing the force, which we have already discussed, it is better to employ the equation in the form in which it is integrated round the circuit s. Thus we have for F

$$F = K \left\{ \sum \frac{i.\mathbf{ds} \wedge r}{r^3} \right\} \wedge i'.\mathbf{ds'}.$$

The constant K has been introduced, as has already been mentioned, to take account of the possibility of employing different systems of units. We have seen that if Ampère's electrodynamic units are employed, $K = \frac{1}{2}$. If electromagnetic units are employed, $K = 1$. In both these cases the force F is measured in dynes. If, on the other hand, the ampere (one-tenth of an electromagnetic unit) is employed as the unit for the measurement of current, the metre (100 cm) for the unit of length and the newton (10^5 dynes) for the unit of force, it is evident that $K = 10^{-7}$.

Thus in the m.k.s. system of units, the force (measured in newtons) is given by the vector product of

$$10^{-7} \left\{ \sum \frac{i.\mathbf{ds} \wedge r}{r^3} \right\} \quad \text{and} \quad i'\mathbf{ds'}.$$

It is usual, in any system of units, to speak as we have done already, of

$$K \left\{ \sum \frac{i.\mathbf{ds} \wedge r}{r^3} \right\}$$

as the magnetic induction produced, at the point occupied by $\mathbf{ds'}$, by the current i in the circuit s, and to represent it by the vector $\mathbf{B}$.

In the m.k.s. system the vector $\mathbf{B}$ in free space, outside magnetic material, is broken down still further. There are two systems—the so-called "rationalized" and the "unrationalized" systems—though it is only a question of convenience and convention between the two. The term originated from Oliver Heaviside. He also had a "rational current element" for which there is a good deal to be said, although the implied irrationality imputed thereby to other concepts, does not necessarily exist.

In the unrationalized system the vector

$$\sum \frac{i.\mathbf{ds} \wedge \mathbf{r}}{r^3}$$

is referred to as the "magnetic intensity" and denoted by $\mathbf{H}$, while the constant $K = 10^{-7}$ is written μ_0 and called the "permeability of free space". Thus in the unrationalized system

$$\mathbf{B} = \mu_0.\mathbf{H} \quad \text{and} \quad \mu_0 = 10^{-7}.$$

In the "rationalized" system the term "magnetic intensity" and the symbol $\mathbf{H}$ is allotted to the quantity

$$\mathbf{H} = \frac{1}{4\pi} . \sum \frac{i.\mathbf{ds} \wedge \mathbf{r}}{r^3} .$$

To compensate, of course, the factor 4π has also to be introduced into the "permeability of free space" so that,

$$\mu_0 = 4\pi.10^{-7}.$$

Again,

$$\mathbf{B} = \mu_0.\mathbf{H}.$$

The attempt to distinguish two different vectors $\mathbf{B}$ and $\mathbf{H}$, in free space has led to a good deal of confusion and seems to serve little purpose. The fundamental fact, deductions from which it is possible to test experimentally, is that electric circuits exert forces on each other which may be calculated by integrating Ampère's law for current elements. It seems reasonable to break the expression down into two parts, corresponding to the two vector products, one being thought of as a field generated by one of the circuits and the other giving the force experienced by the other when placed in it, although even this is really little more than convention. The gain to be obtained by the further decomposition of these constituent quantities is a little difficult to discern.

The case is different when magnetic media have to be considered. Here two vectors may be distinguished experimentally.

Even here, however, it is well to keep the two the same dimensionally. We can return to this point after we have considered theories of magnetism to which Ampère's paper will lead us. In the meantime a word about dimensions generally may not be out of place.

It is usual to refer the dimensions of a physical quantity to the three fundamental dimensions of mass, length and time. For example, from Newton's definition making force proportional to mass times acceleration, or the equation

$$F = K.M.a,$$

it follows that the dimensions of force must be

$$M\,L\,T^{-2}$$

if K is to be a dimensionless quantity. In this case there is no good reason for assigning dimensions to K but the decision is to a considerable extent arbitrary. If we did decide to assign dimensions to K we could alter the dimensions of force in any way we pleased. Alternatively we could add force to our list of fundamental dimensions and say that K has the dimensions of

$$F\,M^{-1}\,L^{-1}\,T^{2}.$$

The unit of current may be obtained by measuring the force between two circuits, the action of which we can calculate by means of Ampère's law of action of current elements. For example, we might measure the force per length ds between two straight wires carrying the same current i and of infinite length, at a distance d apart. By Ampère's law the force would be,

$$F = K\,\frac{2.i^{2}.ds}{d},$$

If we treat the constant K here, in the same manner as we did the similar constant in the relation defining force, and make it dimensionless, it would then follow that the dimensions of current must be

$$M^{\frac{1}{2}}L^{\frac{1}{2}}T^{-1}$$

On the other hand, we may, if we wish, introduce a new fundamental dimension, electric current I into our list, in which case the dimensions of K become,

$$M\,L\,T^{-2}\,I^{-2}.$$

Considerations such as these affect the dimensions to be assigned to μ_0, B and H on the m.k.s. system. Thus if we take the rationalized system, we have for the force between current elements in free space,

$$F = \frac{\mu_0}{4\pi} \left\{ \sum \frac{i.\mathbf{ds} \wedge \mathbf{r}}{r^3} \right\} \wedge i'.\mathbf{ds}'$$

and for H

$$H = \frac{1}{4\pi} \left\{ \sum \frac{i.\mathbf{ds} \wedge \mathbf{r}}{r^3} \right\}$$

and for B

$$B = \mu_0.H.$$

Including electric current among our list of fundamental dimensions we have for the dimensions of μ_0

$$M L T^{-2} I^{-2}.$$

The dimensions of H are,

$$L^{-1} I$$

and of B, are

$$M T^{-2} I^{-1}.$$

If we adopt this system of conventions it follows that we must have B and H of different dimensions and, at the same time, μ_0 cannot be a dimensionless constant. On the other hand, it is clearly a mistake to associate any inner meaning to this state of affairs. The dimensions of μ_0 have to do with the "properties of free space" only in so far as these properties are assigned to it by the system of conventions which we have decided to adopt. Conventions may be made to suit our convenience and, if our convenience is really served by them, there cannot be the slightest reason why they should not be employed. We need, however, to count the gains very carefully before complicating our system by multiplying the entities with which we have to deal. In developing any field theory we have already introduced one mental concept between us and the facts which are open to observation. Clearly we should think carefully before introducing two in a place where one will serve.

There is a further consideration which might conveniently be taken at this stage. If we write q for electrical charge and thus q/t for current, then it follows from the equation for the force between current elements that

$$\mu_0.q^2.T^{-2}$$

must have the dimensions of force.

If we now turn to electrostatics and employ the equation for the force between two electrical charges in empty space (again in rationalized m.k.s. units)

$$F = \frac{1}{4\pi\varepsilon_0} \cdot \frac{q.q'}{r^2} \, ,$$

we see that $\varepsilon_0^{-1}.q^2.L^{-2}$ must also possess the dimensions of force. It follows, in consequence, that $1/(\varepsilon_0\mu_0)$ must have the dimensions of $L^2 T^{-2}$, that is, of the square of a velocity. It turns out, of course, on the electromagnetic theory that $\varepsilon_0\mu_0$ is equal to $1/c^2$, where c is the velocity of light. The product $\varepsilon_0\mu_0$ is, therefore, an experimentally observed quantity and cannot be fixed arbitrarily. On the rationalized m.k.s. system of units μ_0 is given the arbitrary value of $4\pi \times 10^{-7}$ kilogram.metre.second^{-2}.ampere^{-2}, or as it is usually expressed, henry per metre, and this is obviously not an experimental number. In consequence, the value for ε_0 on this system,

$$\varepsilon_0 = 8{\cdot}85 \times 10^{-12} \text{ kg}^{-1} \text{ metre}^{-3} \text{ sec}^4 \text{ amp}^2$$

or farads per metre, must be an experimental quantity. It is determinable, in principle, by measuring the force between two charges, each of one coulomb, separated by a distance of one metre.

At the moment we cannot carry this discussion any further until we have examined what happens within magnetic material. The basis for this examination will be furnished by Ampère's theory of magnetism to which we must now turn.

G.   AMPÉRE'S QUANTITATIVE THEORY OF MAGNETISM

It remains, now, to consider Ampère's theory of magnetism which he set out quantitatively in the same Mémoire in which he gave the theory of the law of action of current elements. The qualitative description of the theory of magnetism is described in the Mémoire previously quoted (Part 2, Section 3) and it is easy enough to appreciate. It is simply that magnets may be looked on as assemblies of particles, all pointing in the same direction, round which circulate perpetual electric currents. The Mémoire of 1823, from which the extracts in Part 2, Section 4G are taken, deals with the mathematical development of this idea. In it Ampère shows that his law of action of current elements accounts satisfactorily for the fact that the forces experienced by a magnet are directed towards two points or poles, one at each end of the magnet, and quantitatively for Coulomb's law of inverse squares, for the forces between poles. He also shows that although the forces experienced by elements of an electric current in the neighbourhood of a magnet are at right angles to the element, the force experienced by the pole itself passes through the pole. The force between two magnetic poles is also shown, according to the theory, to be along the line joining the poles, and he thus succeeded in unifying the phenomena in the two fields of magnetism and current electricity in detail.

It will prove possible to condense the calculations which Ampère gave in his Mémoire by using the results which we have already arrived at, but first it is necessary to extend the theory we have already considered so as to include the deduction of certain expressions which he used in his subsequent theory of magnetism. To follow the mathematical theory which he developed on the basis of his conception of magnetism we shall wish to take up the argument at the point where he discussed the action of a long solenoid of small cross-section, upon which his theory of magnetism is founded. As an element in such a solenoid he considers a small closed circuit of area λ, and we can arrive at the formulae from which he starts in this part of his Mémoire from very simple considerations.

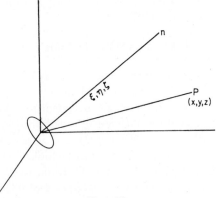

FIG. 22.

Let us take, with Ampère, the origin of coordinates at the centre of the small circuit and let the angles which the normal to the area of the circuit makes with the three axes be ξ, η, ζ. We consider the field generated at the point $P(x, y, z)$ in (Fig. 22).

The angle θ between OP and the normal to the circuit is given by

$$\cos \theta = \frac{x}{l} \cdot \cos \xi + \frac{y}{l} \cdot \cos \eta + \frac{z}{l} \cdot \cos \zeta.$$

The solid angle ω subtended by the circuit at P is

$$\omega = \frac{\lambda \cos \theta}{l^2}$$

$$= \lambda \left\{ \frac{x \cos \xi}{l^3} + \frac{y \cos \eta}{l^3} + \frac{z \cos \zeta}{l^3} \right\}.$$

But

$$l = \sqrt{(x^2 + y^2 + z^2)} \quad \text{and} \quad \frac{dl}{dx} = \frac{x}{l}.$$

Therefore,

$$\frac{1}{K.i} B_x = \frac{d\omega}{dx} = \frac{-3\lambda.x(x\cos\xi + y\cos\eta + z\cos\zeta)}{l^5} + \frac{\lambda\cos\xi}{l^3}$$

$$= \frac{-3\lambda x.l\cos\theta}{l^5} + \frac{\lambda\cos\xi}{l^3}.$$

Now $l \cos \theta$ is the projection of the radius vector l on to the normal to the circuit and is called q by Ampère. This gives us the value of his first integral

$$A = \lambda \left\{ \frac{\cos \xi}{l^3} - \frac{3qx}{l^5} \right\}.$$

In the paragraphs which follow, Ampère then integrates this and similar expressions for B_y and B_z for one of his elementary solenoids.

We are now in a position to follow Ampère's calculations directed towards his theory of magnetism. After this comes a long section of heavy mathematical work in which the theory of the equivalent magnetic shell is worked out for a system of closed circuits. The forces experienced by the circuits are shown to be identical to those which would be experienced by the equivalent magnetic shells with which Ampère is thus able to replace them. Apart from the admiration for the mathematical genius of Ampère which it engenders, there is little point in following this argument through in detail. In the form in which Ampère gave it, it adds little to the results already obtained in terms of current elements to which it is, of course, entirely equivalent. In fact it represents a movement in a direction counter to that taken by him in the general lines of the remainder of his paper. This was, of course, that magnetism should be explained in terms of electric currents, and not vice versa, as this passage seems to imply. On the basis of our potential function proportional to the solid angle subtended by a circuit, as already described in these comments, all the results of the theory of magnetism in the form of magnetic shells becomes available to us, but without the necessity for picturing any distribution of magnetism If Ampère's thesis is to be accepted then we need take account of nothing but electric currents and we need not bother our heads with distributions of magnet poles.

The potential function proportional to solid angle enables us to simplify, very considerably, Ampère's work leading to the theory of magnetism. Let us consider the magnetic induction developed by one of his elementary solenoids, infinite in length in one direction.

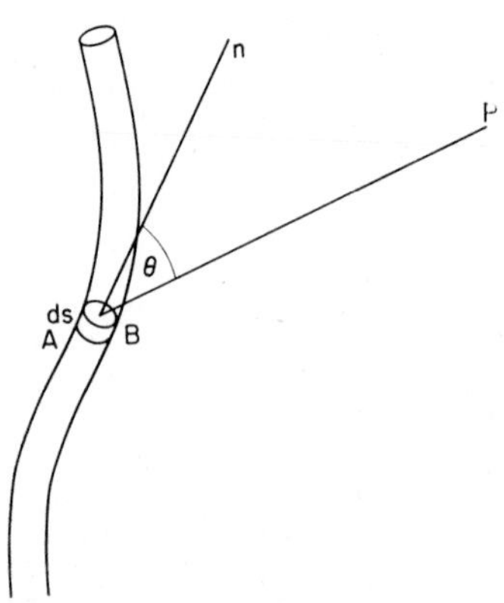

FIG. 23.

Consider the section AB, of thickness ds, of such a solenoid (Fig. 23). Let us write with Ampère, the current flowing round its circumference as ds/g,[†] except that we will not employ Ampère's electrodynamic units. Thus $1/g$ is the current circulating per unit length of the solenoid in any convenient unit.

The magnetic potential at P arising from AB, will therefore be

$$K. \frac{\omega.ds}{g},$$

ω being the solid angle subtended by the section at P. Thus the magnetic potential due to AB is

$$K \frac{\lambda \cos\theta}{r^2} \frac{ds}{g}.$$

Since $\cos\theta = dr/ds$, we have for the magnetic potential which is due to the elementary section of the solenoid,

$$K \frac{d}{ds} \left\{ \frac{\lambda}{g} . \frac{1}{r} \right\} ds.$$

That which is due to the whole solenoid is obtained by integrating this expression along its length—that is with regard to ds. It becomes

$$K \frac{\lambda}{g} \left\{ \frac{1}{r_1} - \frac{1}{r_2} \right\},$$

where r_1 and r_2 are the lines from P to the two ends of the solenoid. If the solenoid is infinite in length in one direction, like those considered by Ampère, the expression for the magnetic potential is simply

$$K \frac{\lambda}{g} . \frac{1}{r},$$

where r is now the distance of P from the near end.

The field of magnetic induction generated by the solenoid is the gradient of this quantity. It is obviously directed radially outwards from the end of the solenoid and varies inversely as the square of the

---

[†] Ampère actually writes this $i. \dfrac{ds}{g}$.

distance from it. This part of the theory thus clearly shows that the familiar behaviour of a long magnet will be mirrored by the action of the long solenoid so far as its action on a current element or on a collection of current elements is concerned. By placing a number of such solenoids side by side the behaviour of magnets of finite cross-section is also explained. Furthermore, as Ampère himself explained, if the solenoids are not completely lined up, it is possible to account for the fact that the poles of a magnet do not reside strictly at the ends.

The theory based on the mutual action of current elements has thus so far shown that it can account for the mutual action between two or more electric currents and the force exerted by a magnet on an electric current. To complete the picture it remains to show that the action of one magnet on another can also be accounted for in the same way. To do this we must examine the force exerted by one solenoid on another.

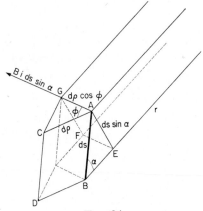

FIG. 24.

We first note that if a current element $i.\mathbf{ds}$ lies in a field of magnetic induction $\mathbf{B}$ along the direction of r, making an angle α with it, as in Fig. 24, the force it experiences is $B.i.ds \sin \alpha$ in a direction perpendicular to the plane of $\mathbf{ds}$ and $\mathbf{B}$, and that the component of the force in a direction making an angle φ with it will be $B.i.ds \sin \alpha \cos \varphi$. The work done by the force when the element is displaced a distance $d\rho$ in this direction will be $B.i.ds \sin \alpha \cos \varphi.d\rho$.

Figure 24 is geometrically identical to Fig. 19. The quantity ds.dφ sin α cos φ is equal to the area *AEFG*, so that B.ds.dφ sin α. cos φ is equal to δN, the number of lines of induction cut by the element during the displacement dφ from *AB* to *CD*.

In the previous discussion based on Fig. 19, when summation over a complete circuit was performed, each element of the circuit received the same displacement. When we sum the results for the elements of a complete circuit in the present case, however, this restriction is no longer necessary. The result we are using is true for each element and not only when summed for a complete circuit; the work would actually have to be performed on each element. It is true that, if we accept Ampère's fundamental assumptions, we have to restrict the application of the Biot–Savart formula to complete circuits, but this we have already done in the present case since we are assuming that B is already given. The function of the Biot–Savart formula is to enable us to calculate B, and in doing this we must, according to Ampère, always integrate over a complete circuit. Here we are assuming that this has already been done. By allowing dφ to vary from one part of a circuit to another in the present case we can include, for example, rotations, with or without simultaneous translations, within the scope of the result.

The work done in moving ds from *AB* to *CD*, is thus $i.\delta N$. Summing this for a complete circuit, we see that the work done in moving the whole circuit from one position to a neighbouring one is $i.dN$, where dN is the change in the number of lines of induction embraced by the circuit.

The result can obviously be integrated so as to apply to displacements of finite magnitude, so that if a circuit is moved from a position where it embraces N_1 lines of induction to another where it embraces N_2, the work which must be performed will be

$$i\,(N_2 - N_1).$$

Throughout this displacement it is assumed that not only is the current in the circuit i maintained constant, but that this applies also to the current in the circuit or system of circuits which generate the field of induction B. In general this will entail continual adjustment of both circuits.

Let us now use this result to show that the line of action of the force experienced by a solenoid, infinite in length in one direction, when placed in the neighbourhood of another, passes through its extremity. Consider the action of one semi-infinite solenoid PQ

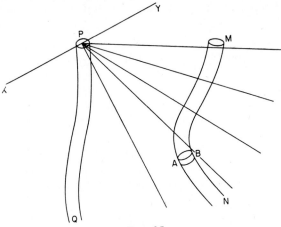

Fig. 25.

(Fig. 25), terminating at P on an element of another, MN, which terminates at M nearby. We have seen that the lines of induction produced by the solenoid PQ radiate from the end P. A certain number of these will thread the elementary circuit AB forming part of the second solenoid MN.

Consider the displacement of AB caused by a rotation about any line passing through P. Since the lines of induction radiate from P, this will cause no change in the number of these lines which thread AB. The forces experienced by AB, therefore, will perform no work during this displacement and consequently can possess no moment about this line XY. They must, therefore, reduce either to a single force which passes through XY or to a couple in a plane containing XY. But XY is any line through P and it therefore follows that the forces experienced by AB must reduce to a single force which always passes through XY in whichever direction it is drawn. The resultant must in consequence pass through P.

This being true for all elements of the solenoid MN, the total resultant force experienced by this solenoid in the neighbourhood of

PQ, must also pass through P. It is a simple matter to calculate its magnitude and direction.

We have already seen that the magnetic potential to which a semi-infinite solenoid gives rise is given by the expression

$$K \frac{\lambda}{g} \cdot \frac{1}{r},$$

where λ is the area of cross-section of the solenoid, $1/g$ is the current measured in suitable units which circulates round it per unit length, and r is the distance from its end. The magnetic induction, which is the gradient of this quantity, will therefore be

$$B = -K \frac{\lambda}{g} \cdot \frac{1}{r^2}$$

in magnitude and it will be directed radially outwards from the end. Let λ and g refer to the solenoid PQ (Fig. 25) and let λ' and g' represent similar quantities for the solenoid MN.

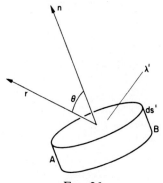

Fig. 26.

The number of lines of induction from the solenoid PQ which thread the elementary circuit AB forming part of MN (Fig. 26) will be

$$N = K \frac{\lambda}{g} \cdot \frac{1}{r^2} \lambda' \cos \theta,$$

where θ is the angle between r and the normal to AB.

The component of the force which is experienced by AB, resolved in any direction ρ, will be

$$\delta F_\rho = \frac{ds'}{g} \cdot \frac{dN}{d\rho}$$

$$= K \frac{\lambda . \lambda'}{g.g'} \cdot \frac{d}{d\rho} \left\{ \frac{\cos \theta}{r^2} \right\} ds'.$$

Thus for the whole solenoid MN, the component of the force which it experiences, resolved in the direction ρ will be

$$F_\rho = K \frac{\lambda . \lambda'}{g.g'} \frac{d}{d\rho} \int \frac{\cos \theta}{r^2} ds'.$$

But

$$\cos \theta = \frac{dr}{ds'}.$$

So that

$$F_\rho = K \frac{\lambda . \lambda}{g.g'} \frac{d}{d\rho} \int \frac{1}{r^2} \cdot dr.$$

$$= K \frac{\lambda . \lambda'}{g.g'} \frac{d}{d\rho} \left\{ \frac{1}{r_1} - \frac{1}{r_2} \right\},$$

where r_1 and r_2 are the distances from P of the two ends of the solenoid MN. If MN is infinite in the direction of N this reduces to

$$F_\rho = K \frac{\lambda . \lambda'}{g.g'} \frac{d}{d\rho} \left\{ \frac{1}{r} \right\},$$

where r is now the distance MP.

The force between the solenoids is thus clearly directed along MP, and it varies inversely as the square of the distance between M and P. It is, in fact, equal to

$$F = K \frac{\lambda . \lambda'}{g.g'} \frac{1}{r^2}.$$

This is the expression deduced by Ampère, apart from the factor K which, in his formula, has the value $\frac{1}{2}$, a value which is introduced through his use of electrodynamic units.

This, in a nutshell, is Ampère's theory of magnetism. To apply it to finite solenoids and thus to magnets of finite lengths, Ampère employed the device of using two semi-infinite solenoids possessing the same axis and with equal currents circulating in opposite directions but with the two solenoids displaced axially relative to each other. In this way the behaviour of magnetic poles is completely simulated by that of Ampère's equivalent solenoids.

Though Ampère would like to go further there is, in fact, no compulsion to look upon his theory as more than an interpretation of magnetic phenomena in terms of the mutual action of electric currents and thus unifying them by means of one system. His theory actually necessitates only the adoption of the principle that magnetic materials behave, when magnetized, *as though* there were electric currents circulating round them, the plane in which the circulation takes place being at right angles to the direction of magnetization. The effects of currents in the interior of substances uniformly magnetized cancel each other out, and it is only on the surface that they give rise to forces on other magnetized bodies or on electric currents. It is of interest to see how far these simple ideas of Ampère can take us. Initially, at any rate, it is not necessary to inquire into the nature of these circulating currents.

If a current i amperes circulates round a plane loop of area α, it will possess a magnetic moment $i.\alpha$—that is to say it requires a couple of this magnitude to maintain it parallel to a field of magnetic induction of unit strength. The component of a magnetization vector in any direction will depend upon the component of the total area of the circulating currents at right angles to this direction. Thus if Σ signifies summation over unit volume,

$$M_x = \Sigma \, i.\alpha_x.$$

If this magnetic moment were produced by a current i_x circulating round the area $dy.dz$ of the elementary parallelopipedon $dx.dy.dz$ (Fig. 27),

$$i_x dy.dz = M_x dx.dy.dz$$

so that

$$i_x = M_x.dx.$$

Thus the current circulating round the elementary prism must be M_x per unit length in the direction of x if it is to generate a magnetic moment equal to M_x per unit of volume.

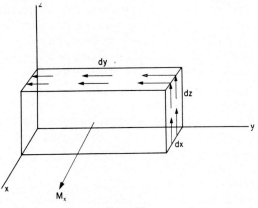

Fig. 27.

To build up larger volumes, elementary prisms may be stacked together. In the case of uniformly magnetized bodies, currents over internal surfaces of neighbouring elements cancel and there is left an external circulating current of M_x per unit length along the axis of x.

When the magnetization is not uniform there will be a current density throughout the material, that in the direction of z being given by

$$j'_z = \frac{\delta M_y}{\delta x} - \frac{\delta M_x}{\delta y}$$

or, more generally,

$$j' = \text{curl } \boldsymbol{M}.$$

If, in addition to the Ampèrian current density $\boldsymbol{j'}$ there is also a "real" current density $\boldsymbol{j}$, the flux of induction will be produced by $\boldsymbol{j} + \boldsymbol{j'}$, so that

$$\text{curl } \boldsymbol{B} = 4\pi K (\boldsymbol{j} + \boldsymbol{j'}) = \mu_0 (\boldsymbol{j} + \boldsymbol{j'}) \quad \text{in m.k.s. units.}$$

That is,

$$\text{curl } (\boldsymbol{B} - \mu_0 \boldsymbol{M}) = \mu_0 \boldsymbol{j}.$$

It is at this point that it is usual to introduce the new vector $H = (B/\mu_0) - M$ known as the magnetic intensity. We have already referred earlier in these comments to the convention of distinguishing between the two vectors B and H in free space (where $M = 0$), differing only in the factor μ_0. Before making further use of the vector H in the case of magnetic bodies, it is desirable to look at the position a little further.

In the interior of atoms and in the interstices between them, a vector such as B will be subject to very large and almost discontinuous variations. Its usefulness in particular cases will depend upon the method of averaging which is employed. To make a magnetic measurement inside a magnetic material it is usually necessary to construct a cavity in it, in which to do so. The results of the measurements are affected by the shape of the cavity constructed, and it is convenient to follow the two well-known cases of the long narrow cavity, on the one hand, and the broad flat cavity, on the other, introduced by Lord Kelvin. Let us try to interpret these two cases on the basis of Ampèrian currents.

For simplicity let us consider a long, uniformly magnetized bar and let us excavate within it a long narrow cavity having its long axis parallel to the axis of the bar which we will also take to be the

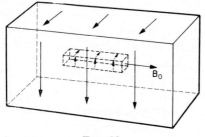

Fig. 28.

direction of magnetization. Ampèrian currents will circulate over the surfaces of the bar as shown in Fig. 28. When the cavity is excavated, however, further Ampèrian currents will develop over the surfaces of the cavity itself and these will circulate in the opposite direction to that of the currents over the surface of the bar. Both sets of currents will contain the same number of amperes per metre in the direction of M. If both bar and cavity are long, points well inside will thus lie within two coaxal solenoids with the same current per unit length. The magnetization of the bar will thus generate no flux of induction at these points.

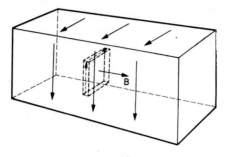

FIG. 29.

If, on the other hand, we measure the magnetic induction in a narrow flat cavity at right angles to the direction of magnetization, there will be no such compensation and the magnetization of the bar will produce its full effect. Since the equivalent Ampèrian current circulating round the bar is M amperes per metre, the magnetic induction will be

$$B = \mu_0 . M.$$

We may note in passing that if we interpret, with Ampère, all phenomena relating to the magnetic effects of steady currents and to magnets, in terms of equivalent circulatory currents, then any measurement made at a point in space can only determine the effects of forces acting upon currents. The results of all determinations must, therefore, be to yield the value of the magnetic induction at the point. No measurement will yield the value of H, the magnetic intensity, which is not directly observable, and can only be obtained indirectly via a determination of B.

In electromagnetic units, in which the constant of proportionality K is equal to unity, the value of the magnetic induction within the long narrow cavity is numerically identical to that of the magnetic intensity H. On the m.k.s. system the two differ by the factor μ_0, to which dimensions are attributed. The problem is one of nomenclature. It would be useful to have a term for the magnetic induction within the long narrow cavity and had the term magnetic intensity not been assigned differently on the m.k.s. system, it would have sufficed to employ it for this purpose. To do so, however, at this stage, would render confusion worse confounded, and the writer has suggested that "basic magnetic induction" and the symbol B_0 might be used for this quantity. It is B_0 rather than H which is involved in all measurements, though the distinction is trivial. B_0 is, of course, the same as $\mu_0 H$.

It is of interest to glance at the variation of the two quantities B and B_0 along the axis of a bar of uniformly magnetized material. In the case of the magnetic induction B, the situation reduces to that of a solenoid. The magnetic induction increases as the bar is approached. It attains half its maximum value at the end of the specimen and increases asymptotically to the maximum value $\mu_0 M$, inside (Fig. 30).

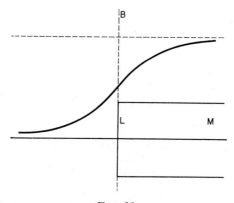

FIG. 30.

The case of the basic magnetic induction B_0, is quite different. In the space outside the bar there is no difference between B_0 and B. At the end of the specimen, however, we enter the long narrow cavity and an opposing magnetic induction, in magnitude equal to the maximum value of B within the bar, is added, since there is no "end effect" in the case of the narrow cavity. At this point B has only attained a value of $\mu_0 M/2$, so that there is a discontinuity at the surface. Further along the axis, within the specimen, B_0 rises to zero (Fig. 31).

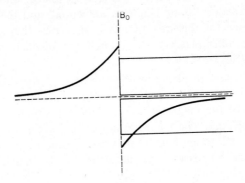

FIG. 31.

The variation in B and B_0 along the axis of a bar magnet may be demonstrated experimentally, using a number of small rectangular magnets of the type used for electrical meters or loud speakers. To measure B, the magnets are taped to a bar of wood, aluminium or other non-magnetic material, leaving narrow gaps between each

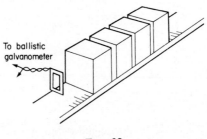

FIG. 32.

unit. The magnetic induction in the gaps may be measured by means of a small, flat search coil connected to a ballistic galvanometer. A diagram of the apparatus is shown in Fig. 32. The results of a series of measurements made by the author are plotted in Fig. 33.

To determine the basic magnetic induction B_0, use is made of small magnets possessing a central fixing hole as in Fig. 34. A small

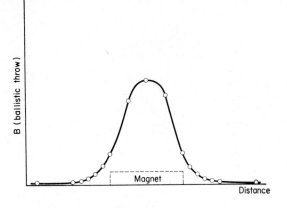

FIG. 33.

search coil of enamelled wire, capable of being passed along the cavity formed by the fixing holes, is wound on an ebonite rod. Readings of the ballistic galvanometer throw, as the coil was moved

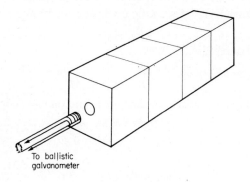

FIG. 34.

from various points on the axis to a large distance away from the magnets, are shown graphically in Fig. 35. The graph follows closely the theoretical predictions of Fig. 31.

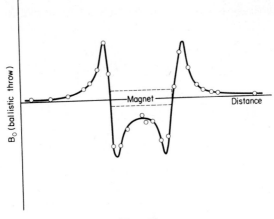

FIG. 35.

We return now to the equation,

$$\text{curl } \boldsymbol{B} = \mu_0 (\boldsymbol{j} + \boldsymbol{j}').$$

In the long narrow cavity, the effect of the local magnetization of the material is neutralized. If we form the integrals of $B_0 ds$ round the elementary rectangles $dx.dy$, $dy.dz$ and $dz.dx$ we see that

$$\text{curl } \boldsymbol{B_0} = \mu_0 . \boldsymbol{j}.$$

We have, therefore,

$$\text{curl } \boldsymbol{B} = \text{curl } \boldsymbol{B_0} + \mu_0 \text{ curl } \boldsymbol{M}$$

or

$$\boldsymbol{B} = \boldsymbol{B_0} + \mu_0 \boldsymbol{M}.$$

For weakly magnetized paramagnetic and diamagnetic substances

$$\mu_0 \boldsymbol{M} = k . \boldsymbol{B_0},$$

where k is called the susceptibility. Also if μ is the permeability of the substance,

$$\boldsymbol{B} = \mu \boldsymbol{B_0}$$

which gives us

$$\mu = 1 + k$$

Let us consider an element of magnetized material in the form of a flat disc of thickness dt, magnetized in the direction of its axis.

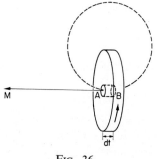

FIG. 36.

If M is the intensity of magnetization, the Ampèrian current circulating round the perimeter will be $M.dt$. Let us consider the integral

$$\int B.ds$$

taken round the dotted curve (Fig. 36) in free space outside the specimen, from a point A just outside on one side to a point B just outside on the other. If the thickness of the element dt is small compared with the diameter, this integral must be equal to $\mu_0.i = \mu_0 M.dt$, since the path is complete except for the small thickness dt. Outside the specimen B and B_0 are identical. Let us complete the path through a narrow cavity along the axis of the disc. Inside, the field will be reversed and the magnetic induction within the material B_0 will be

$$-\mu_0.M.$$

Thus,

$$\int_B^A B_0 ds = -\mu_0.M.dt.$$

Adding this to the value of

$$\int_A^B B_0.ds$$

round the outer path APB we see that when taken round the complete path from A via P to B and back through the specimen to A,

$$\int B_0.ds = 0$$

The Ampèrian currents therefore make no contribution to the integral.

If the path $APBA$ interlaces a "real" current i, in addition, we shall have,

$$\int B_0.ds = \mu_0.i.$$

This agrees with the result that we have already found, namely, that,

$$\text{curl } \boldsymbol{B}_0 = \mu_0.\boldsymbol{j}.$$

As an example of the application of the idea of Ampèrian currents we may consider the magnetic circuit. Let us consider the case of specimens of magnetic materials placed inside a toroidal coil of n turns per unit length carrying a current I. Let N be the total number of turns on the toroid.

Integrating round the line of centres of the cross-section of the toroid we have,

$$\int B_0.ds = \mu_0.N.I.$$

This quantity $\mu_0.N.I.$ is called the magneto-motive force (m.m.f.). If the material has a magnetic permeability μ,

$$\boldsymbol{B} = \mu.\boldsymbol{B}_0,$$

giving

$$\int B_0.ds = \int \frac{B.ds}{\mu}.$$

If φ is the total flux through the specimen and A its area of cross-section, we shall have

$$B = \frac{\varphi}{A}.$$

Thus,

$$\text{m.m.f.} = \int B_0.ds = \int \frac{\varphi.ds.}{\mu A}$$

If we can assume that there is no leakage of flux so that φ remains constant round the toroid, we would have for a series of specimens of different permeabilities, placed end to end inside a toroid,

$$\text{m.m.f.} = \int \frac{\varphi}{\mu A} \cdot ds = \sum \frac{\varphi L}{\mu A},$$

where L is the length of the specimen and the summation extends over all the specimens in the toroid.

Thus we have as the expression for the flux generated

$$\varphi = \frac{\text{m.m.f.}}{\Sigma(L/\mu A)}$$

$$= \frac{\mu_0 . N.I.}{\Sigma(L/\mu A)}.$$

The quantity $L/\mu A$ is called the reluctance of the specimen and thus we see that for a complete magnetic circuit the flux is obtained by dividing the magneto-motive force by the sum of the reluctances of the specimens placed in the toroid.

The principal use of the theory of the magnetic circuit is in connection with ferro-magnetic materials, as in the design of motors and transformers. The value of μ in such a case for iron is likely to be of the order of 2000. Compared with air, therefore, the reluctance of a specimen of iron is likely to be small. Suppose that we have a toroid wound on an iron ring in which there is an air gap of small width. Calling the width of the air gap G, its reluctance will be G/A since μ for air is equal to unity. If we can neglect the reluctance of the iron core in comparison with that of the air gap we shall have for the expression for the flux,

$$\varphi = \frac{\mu_0 . N.I}{G/A}$$

or

$$B = \frac{\varphi}{A}$$

$$= \frac{\mu_0 . N.I}{G}.$$

This is the value for the magnetic induction which would be obtained in a long solenoid wound with a number of turns per unit length equal to the number which would be the case if all the turns on the toroid were concentrated in a length equal to that of the gap.

IV
The Critics

THE theory of the electrodynamics of steady currents developed by Ampère was accepted by his contemporaries immediately. The elegance of its mathematics, the similarity of its methods to those of Newton, the analogy with the theory of gravitation and with the inverse square laws of Coulomb, its acceptance of Newton's laws of motion together with its ability to correlate all the phenomena then known, proved completely convincing. His theory is, indeed, a basic ingredient in modern electrodynamics. Writing in 1873 Maxwell said:

"The experimental investigation by which Ampère established the laws of the mechanical action between electric currents is one of the most brilliant achievements in science. The whole, theory and experiment, seems as if it had leaped, full grown and full armed, from the brain of the 'Newton of Electricity'. It is perfect in form and unassailable in accuracy, and it is summed up in a formula from which all the phenomena may be deduced, and which must always remain the cardinal formula of electro-dynamics."

Nevertheless, Ampère's interpretation of the interaction of two electric circuits, as being the sum of the interactions in pairs of the constituent current elements, each acting upon the other with equal and opposite forces having as their line of action the line joining the elements, has since been criticized by many writers. While it would be understandable and perfectly legitimate to criticize as untestable by experience the basic idea of forces between current elements, most of the criticisms have been levelled against Ampère's particular form of expression. We may here quote Maxwell again, since he made this point very clearly.

"It may be observed with reference to these [Ampère's] experiments that every electric current forms a closed circuit. The currents used by Ampère, being produced by the voltaic battery, were of course in closed circuits. It might be supposed that in the case of the current of discharge of a conductor by a spark we might have a current forming an open finite line, but according to the views of this book [*Treatise on Electricity and Magnetism*] even this case is that of a closed circuit. No experiments on the mutual action of unclosed currents have been made. Hence no statement about the mutual action of two elements of circuits can be said to rest on purely experimental grounds. It is true we may render a portion of a circuit movable, so as to ascertain the action of the other current upon it, but these currents, together with that in the movable portion necessarily form closed circuits, so that the ultimate result of the experiment is the action of one or more closed currents upon the whole or part of a closed current."

Maxwell goes on as follows and gives the correct interpretation to be put upon the conception of the current element:

"In the analysis of the phenomena, however, *we may regard* the action of a closed circuit on an element of itself or of another circuit as the resultant of a number of separate forces, depending upon the separate parts into which the first circuit may be conceived, *for mathematical purposes*, to be divided. This is merely a mathematical analysis of the action, and is therefore perfectly legitimate, whether these forces can really act separately or not".

It is rather remarkable that discussions about the form of the law of action of current elements should have gone on for so long— even down to the present day. The reader may well take the view that questions which, even in principle, cannot be put to the test of experience, do not belong to science, but in spite of this Ampère has been subjected to considerable criticism. Curiously enough, the one criticism to which he would be vulnerable, namely that he, too, along with most of his critics, thought that he was arriving at the law of force which current elements *actually* exerted upon each other, and that he would not have been content with a statement that entire circuits behaved *as if* their constituent elements followed his law, seems rarely to have been put forward.

The first of the critics whom we will notice is Grassmann, and we give his paper in a subsequent section. Hermann Günther Grassmann was born in Stettin on 15 April 1809, and he taught there until his death on 26 September 1877. He developed the idea of using symbols to represent geometrical entities, manipulated according to certain rules, and though, because of the obscurity of his writing his ideas were slow in being adopted, he is looked upon as the father of vector analysis. It is for that reason that the vector expression of Ampère's law is usually known under his name. His criticism of Ampère's expression, however, is not really very profound, and we print his paper as an example of a criticism to which Ampère himself would undoubtedly have reacted with an argument of a metaphysical flavour rather than because it made any significant contribution to the science of electrodynamics which, in fact, it did not do. Grassmann recognised the correctness of Ampère's expression for closed circuits, and based his own work on this assumption. His contribution consisted in providing a more elegant expression, agreeing with Ampère's for closed circuits, and only differing from it for limited circuits, which, as he pointed out, had never been experimented with, though clearly, he thought such tests to be possible.

Grassmann made two points. Of these, one is perfectly valid but inconclusive, and the other falls outside the field of science and is metaphysical. His first point is that current elements possess direction as well as magnitude and that there is, therefore, no *a priori* reason why they should act according to the laws followed by material particles. There would, indeed, have been no *a priori* reason why they should do so even if they were simple scalar quantities and not vector in nature, since they are certainly not material particles. Only experience can supply answers to questions about how they react. Ampère's answer to this criticism of Grassmann would be perfectly clear. In the event it proved possible to describe the mutual action of electric circuits by a simple law, analogous to that between material particles. and there is no point in complicating the theory unnecessarily.

Grassmann's second point, the one which is metaphysical in character, concerns two current elements which are parallel to each other. Ampère's formula shows that the force between them is zero when they make angles of $\frac{1}{2} \cos^{-1}\frac{1}{3}$ with the line joining them, and that it changes from an attraction to a repulsion as the elements are turned round together and their angle with the line which joins them passes through this value. For Grassmann this is too improbable to be acceptable. However, this criticism need occasion no alarm for Ampère. Grassmann can bring no experimental evidence whatever to support his view and there is not the slightest reason to suppose that nature was designed to satisfy the particular tastes of anybody.

Another to be numbered among the critics of Ampère is Oliver

Heaviside, that very original if rather forceful writer. He, indeed, raised objections to the whole idea of current elements as conceived by Ampère. Heaviside was a thorough-going Maxwellian and placed great emphasis upon the importance of the circuital nature of the electric current. On page 64 of his *Electrical Theory* he wrote:

"The two laws of circulation did not start into full activity all at once. On the contrary, although they express the fundamental electromagnetic principles concerned in the most concise and clear manner, it was comparatively late in the history of electromagnetism that they became clearly recognized and explicitly formularized. We have not here, however, to do the work of the electrical Todhunter, but only to notice a few points of interest.

"The first law [this is the so-called 'work rule', namely that $\oint H.ds =$ the current enclosed—in suitable units] had its beginning in the discovery of Oersted that the electric conflict acted in a revolving manner, and in the almost simultaneous remarkable investigations of Ampère. It did not, however, receive the above used form of expression. In fact, in the long series of investigations in electrodynamics to which Oersted's discovery and the work of Ampère, Henry and Faraday, gave rise, it was customary to consider an element of a conduction current as generating a certain field of magnetic force. Natural as this course may have seemed, it was an unfortunate one, for it left the question of the closure of the current open; and it is quite easy to see now that this alone constituted a great hindrance to progress. But so far as closed currents are concerned, in a medium of uniform inductivity, this way of regarding the relation between current and magnetic force gives equivalent results to those obtained from the first law of circulation in the limited form suitable to the circumstances stated.

"If C is the density of conduction current at any place, the corresponding field of magnetic force is given by

$$H = \frac{VC.r_1}{4\pi r^2} \tag{1}$$

at a distance r from the current element C, if r_1 be a unit vector along r from the element to the point where H is reckoned . . . But from the Maxwellian point of view this field of H is that corresponding to a certain circuital distribution of electric current, of which the current element mentioned is only a part; this complete current being related to the current element in the same way as the induction of an elementary magnet is to the intensity of magnetization of the latter. Calling the complete system of electric current a rational current element it may easily be seen that in a circuital distribution of rational current elements the external portion of the current disappears by mutual cancelling, and there is left only the circuital current made up of the elements in the older sense.

We may, therefore, employ the formula (1) to calculate without ambiguity
the magnetic force of any circuital distribution of current. This applies
not merely to the conduction current (which is all that the older electricians
reckoned) but to electric currents in the wider sense introduced by
Maxwell.

". . . . But this method of mounting from current to magnetic force
(or equivalent methods employing potentials) is quite unsuitable to the
treatment of electromagnetic waves, and is then usually of a quite un-
practical nature."

This passage may, perhaps, become clearer if a little more of the
work of "the electrical Todhunter" is attempted. Heaviside bases
his theory of electricity on certain circuital axioms of which the one
with which we are concerned here is the "work law". His exposition,
like that of Newton and Ampère, is Euclidian in character, the aim
being to produce a logically coherent scheme based upon a concise
system of axioms. The origin of the axioms in experience is not
discussed, this presumably being left to a process of verification
a posteriori.

Heaviside's so-called rational current element comprises a flow
of current i along the element AB (Fig. 37) "in the older sense",

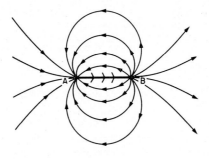

FIG. 37.

the circuital current being completed in the surrounding space along
the lines of force of a magnet coinciding with AB. This spatial return
flow may be looked upon as arising from a sink of strength i placed
at A and a source of equal strength at B. When such current elements
are placed end to end, it is obvious that these spatial go and return
currents from neighbouring elements, will cancel, as Heaviside

points out. So long as we deal with closed circuits, therefore, no change is introduced by this innovation.

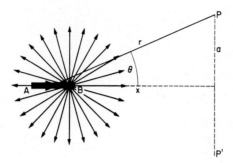

Fig. 38.

FIG. 38.

Consider the field generated by the Heaviside current element AB at the point P (Fig. 38). By symmetry it will be the same all round the circumference of the circle PP', whose axis is AB and radius a. Let the solid angle subtended by this circle at A and B respectively be ω_1 and ω_2. The current which flows through the circle into A will be $\frac{i}{4\pi} \cdot \omega_1$, and that which flows through it in the opposite direction out of B will be $\frac{i}{4\pi} \cdot \omega_2$. The net current flowing through the circle will, therefore, be

$$\frac{i}{4\pi} (\omega_1 - \omega_2) = \frac{i}{4\pi} \cdot \frac{d\omega}{dx} \cdot ds$$

$$= \frac{i}{4\pi} \cdot \frac{d}{dx} \left\{ 2\pi (1-\cos\theta) \right\} ds$$

$$= \frac{i}{2} \cdot \frac{d}{d\theta} \left\{ 1-\cos\theta \right\} \cdot \frac{\sin\theta}{r} \cdot ds$$

$$= \frac{i.}{2} \cdot \frac{\sin^2\theta}{r} \cdot ds.$$

Applying the work law and using H to represent the magnetic field in conformity with Heaviside's notation, we have,

$$2\pi.a.H = \mu_0 \frac{i}{2} \cdot \frac{\sin^2 \theta}{r} \cdot \mathrm{d}s$$

which gives

$$H = \mu_0 \frac{i.\mathrm{d}s \sin \theta}{4\pi r^2}$$

which is equivalent to Heaviside's equation (1).

Since, as we have seen, the work rule is derivable from Ampère's law of action of current elements, all that this deduction tells us is that a Heaviside current element, which is conformable to the Maxwellian idea of current as a continuous circulation, follows the Biot–Savart form of Ampère's law.

We need to make some further comment on Heaviside's last paragraph, since mounting from current to magnetic force is precisely what we are engaged upon in this book. In spite of what he says, of course, he does so too. However, he conceals the fact by starting off with the circuital laws which he takes for his axioms. The origin of these in the observation of the behaviour of electric currents in wire circuits he dismisses in a sentence or two.

To clear this point up it is well to distinguish between logical priority and epistemological priority. Epistemologically the properties of currents in wire circuits were known before the axioms with which Heaviside begins. Nevertheless, in his development they figure as deductions. He appears to present the deduction as a justification for this prior knowledge. What he is doing, in fact, is to develop a self-contained logical system. In this system the older current elements follow and do not precede his axioms. The purpose of the demonstration, therefore, can only be to justify the generalization by showing its conformity with what is already known to be the case. By uniting these pieces of existing knowledge to new facts which are deducible from his more general theory he adds something, though perhaps not a very great deal in this case, to the certainty with which they are known. The most accurate confirmation of Ampère's law is probably to be found in the agreement obtained in the use of current balances in the measurement of current. It is direct and of an accuracy of the order of 1 in 10^5.

Later still, Whittaker† rejected the Biot–Savart and Grassmann formulae on the ground that they were not in accordance with Newton's third law. With this, of course, Ampère would find himself in warm agreement. But Whittaker also rejected Ampère's own expression as well, on the ground that it neglected the possibility of current elements exerting a turning effect upon each other in addition to simple attraction and repulsion. The point is fundamentally the same as that raised by Grassmann, but Whittaker proceeds to give a more general expression. Ampère was certainly aware of the possibility of turning effects since he refers to it in the case of magnetic molecules. As we have already seen, it is true that his choice of simple attraction and repulsion as the mode of action of current elements was based upon theoretical notions about what constituted a "truly elementary force". But having succeeded in solving the problem on this basis, he then is equipped with the perfectly adequate defence that further complications are unnecessary. His solution is the simplest. Moreover, this defence can be put on solid, pragmatic grounds. It is not necessary for him to invoke any metaphysical principle about the simplicity of nature. There is no reason to suppose that nature corresponds to the simplest structure of which we can conceive. It is, however, surely only sensible not to complicate our calculations unnecessarily and, so far as is known, the assumption that steady electric currents in closed circuits behave *as if* their constituent elements obeyed Ampère's law (in whatever form we choose to employ it) is perfectly adequate to describe the phenomena.

It might be thought, at first sight, that it would be possible to distinguish between the Ampèrian and the Biot–Savart expressions by measurements of the force which a part of a circuit exerts upon the remainder, as in a circuit breaker, for example. Indeed, 100 years after Ampère, experiments to this end were attempted and led to some discussion spread over about 10 years from 1936.‡ The

---

†*Theories of the Aether and Electricity*, **1**, 87.
‡F. F. Cleveland, *Phil. Mag.* **22**, 416 (1936).   Dunton, *Nature* **140**, 245, (1937).   Mathur, *Phil. Mag.* **32**, 171 (1941).   Robertson, *Phil. Mag.* **36**, 32, (1945).

situation is not resolvable in this manner since, if the circuit is assumed to consist of line currents the integrals to which both formulae lead become infinite. If, to get over this difficulty, the circuit is divided into constituent elementary filaments then, in effect, we calculate the force experienced by each arising from the remainder, which form closed circuits, so that again both formulae produce identical results. Time is not profitably spent in discussing experiments of this type, but it is of interest to see how the situation arises.

To see that the integrals become infinite when the currents are concentrated in a line, let us try to calculate the force experienced by an element ds of a circular current arising from the remainder of the circuit.

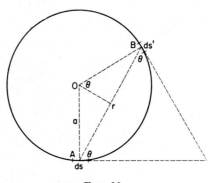

FIG. 39.

First, according to the Biot–Savart formula, the force on ds at A (Fig. 39) arising from ds' at B will be along the radius OA and equal in magnitude to

$$\frac{i^2 ds ds' \sin \theta}{r^2} = \frac{i^2 ds \, d\theta}{2a \sin \theta}$$

The total force on ds is

$$\frac{i^2 ds}{2a} \int_0^\pi \operatorname{cosec} \theta \, d\theta = \frac{i^2 ds}{2a} \left[\ln \tan \frac{\theta}{2} \right]_0^\pi$$

which is infinite.

Similarly, Ampère's expression gives for the force along OA

$$i^2 ds \int_0^\pi \frac{(\cos^2 \theta + 2\sin^2 \theta) \sin \theta}{2a \sin^2 \theta} \, d\theta$$

$$= \frac{i^2 ds}{2a} \int_0^\pi (\operatorname{cosec} \theta + \sin \theta) \, d\theta$$

which is also infinite.

Both formulae, therefore, indicate that it is necessary to take into account the finite thickness of the wire. Before leaving this question let us actually calculate the force on an element of a circular circuit made of wire of radius r, bent into a circle of radius a.

FIG. 40.

Consider the mutual action of two filaments F_1F_1' and F_2F_2' of radius a_1 and a_2 respectively (Fig. 40). Let the number of lines of induction threading F_1F_1' arising from F_2F_2' be n_1. Let the radial force experienced by F_1F_1' per unit length be p_1. If we allow F_1F_1' to expand while keeping F_2F_2' fixed we shall have

$$2\pi a_1 p_1 = j \, \frac{\partial n_1}{\partial a_1} \, d\alpha_1,$$

where $d\alpha_1$ is the cross-section of the filament F_1F_1' and j is the current density.

Similarly, the radial force experienced per unit length by F_2F_2' will be given by,

$$2\pi a_2 p_2 = j \, \frac{\partial n_2}{\partial a_2} \, d\alpha_2.$$

If m is the coefficient of mutual induction between the two filaments,
$$n_1 = mj\mathrm{d}\alpha_2 \text{ and } n_2 = mj\mathrm{d}\alpha_1.$$
So that if both filaments are allowed to expand by the same amount
$\mathrm{d}a$,

$$2\pi\,(a_1p_1 + a_2p_2) = j^2\mathrm{d}\alpha_1\mathrm{d}\alpha_2 \left\{ \frac{\partial m}{\partial a_1} + \frac{\partial m}{\partial a_2} \right\}$$

$$= j^2\mathrm{d}\alpha_1\mathrm{d}\alpha_2\,\frac{\mathrm{d}m}{\mathrm{d}a}.$$

If P is the radial force per unit length experienced by the whole
wire,
$$2\pi a P = \Sigma 2\pi a_1 p_1$$
$$= \tfrac{1}{2}\,\Sigma\,2\pi\,(a_1p_1 + a_2p_2),$$

since in the last expression each filament would be counted twice
over.

So that

$$\left.\begin{aligned}
2\pi a P &= \frac{j^2}{2}\,\Sigma\,\mathrm{d}\alpha_1\mathrm{d}\alpha_2\,\frac{\mathrm{d}m}{\mathrm{d}a} \\[2mm]
&= \frac{i^2}{2}\,\frac{\mathrm{d}L}{\mathrm{d}a}
\end{aligned}\right\} \tag{1}$$

where L is the coefficient of self-induction for the circle of wire and
i the total current. If r is the radius of the cross section of the
wire, to the first order in r/a the value of L is

$$L = 4\pi a \left\{ \ln\frac{8a}{r} - \frac{7}{4} \right\}$$

which gives

$$P = \frac{i^2}{a} \left\{ \ln\frac{8a}{r} - \frac{3}{4} \right\}.$$

Equation (1) can be obtained directly by invoking the conservation
of energy. It is necessary to be careful how this principle is applied
to electric circuits. If they contain batteries, energy can be derived
from them and it is not possible to say, *a priori*, that external work

must balance changes in electro-kinetic energy. The above calculation was dependent only upon Ampère's law. This book is limited to considerations of steady currents and the concept of electro-kinetic energy, which arises from the work of Faraday is, strictly speaking, excluded. If we may overstep these limits for a moment we can obviate the difficulty of the battery by considering a ring of a superconductor in which a current will flow without an external source of energy. In such a circuit no e.m.f. can be generated. Thus,

$$\frac{d}{dt} (L i) = 0$$

giving

$$L \frac{di}{dt} = - i \frac{dL}{dt}$$

When a ring of such material is expanded, the work performed must equal the change in the energy of the field. Thus,

$$2\pi a P = - \frac{d}{da} \left\{ \frac{L i^2}{2} \right\} = - \left\{ \tfrac{1}{2}i^2 \frac{dL}{da} + L i \frac{di}{da} \right\}$$

$$= \frac{i^2 dL}{2 \, da}$$

which is equation (1). The method is of general application.

There is one further type of criticism which has recently been levelled against Ampère's work, at which we must look in closing. This again is strictly outside the limits of this book, since it involves current elements which radiate and are, thus, not steady. Electro-dynamics has two roots. One is in the work of Ampère, who dealt with steady currents. From it arises what Heaviside called the first law of circulation. The second root is in the work of Faraday, which is concerned with varying currents and leads to the second law of circulation. If one is to do more than be able to apply, more or less blindly, a deductive system based upon far from self-obvious axioms, it is well to appreciate the origin of these roots. It has been the object of this book to trace the development and some of the consequences of the first law of circulation.

The last type of criticism which we shall consider has been put forward by Geoffrey Builder† and by Rosser.‡ Both take as their current element a flow of negative electric charges past stationary positive charges in a small element of conductor. Starting with the Coulomb law of inverse squares between static charges, they apply the relativity formulae to the case where the charges move and arrive, as a result, at the Biot–Savart expression for the mutual action between such elements, to the first order in v/c. As a result Builder declared the Ampèrian formula to have been disproved as being incompatible with the restricted theory of relativity and the accepted electrodynamical theory of electrons.

When isolated, current elements of the kind considered, radiate energy. Although at the instant $t=0$ the charges of opposite sign coincide, this is not true as time progresses. An electric dipole is generated as the charges separate and this gives rise to radiation. The momentum carried away in the radiation has been shown by Page and Adams* to balance the difference in the forces which two current elements exert upon each other. Rosser stresses that the current elements should not be thought of as completely isolated from their circuits. In this he is quite correct. In a complete circuit, electric dipoles are prevented from developing by mutual cancellation and there is no radiation. But so long as current elements are maintained in complete circuits it is impossible to distinguish between Ampère's expression and that of Biot and Savart. For complete circuits both agree with or infringe the principle of relativity together. The upshot of these discussions must, therefore, be simply to imply that the law of action of current elements should be used only in an integrated form. An isolated element of a steady current is a contradiction in terms. Nevertheless, as Maxwell said, it is perfectly legitimate to conceive of circuits being divided into elements each exerting separate forces upon each other, for purposes of mathematical calculation, whether these forces can really act separately or not.

---

† Builder, *Bull. Inst. Phys.* 9, 12 (1958).
‡ W. G. V. Rosser, *Contemporary Physics*, Vol. 3, 1961.
* Page and Adams, *Amer. J. Phys.* 13, 141 (1945).

Part 2

Part 2

1

A Oersted

EXPERIMENTS ON THE EFFECT OF A CURRENT OF ELECTRICITY ON THE MAGNETIC NEEDLE*

THE first experiments respecting the subject which I mean at present to explain, were made by me last winter, while lecturing on electricity, galvanism, and magnetism, in the University. It seemed demonstrated by these experiments that the magnetic needle was moved from its position by the galvanic apparatus, but that the galvanic circle must be complete, and not open, which last method was tried in vain some years ago by very celebrated philosophers. But as these experiments were made with a feeble apparatus, and were not, therefore, sufficiently conclusive, considering the importance of the subject, I associated myself with my friend Esmarck to repeat and extend them by means of a very powerful galvanic battery, provided by us in common. Mr. Wleugel, a Knight of the Order of Danneborg, and at the head of the Pilots, was present at, and assisted in, the experiments. There were present likewise Mr. Hauch, a man very well skilled in the Natural Sciences, Mr. Reinhardt, Professor of Natural History, Mr. Jacobsen, Professor of Medicine, and that very skilful chemist, Mr. Zeise, Doctor of Philosophy. I had often made experiments by myself; but every fact which I had observed was repeated in the presence of these gentlemen.

The galvanic apparatus which we employed consists of twenty copper troughs, the length and height of each of which was 12 in.; but the breadth scarcely exceeded $2\frac{1}{2}$ in. Every trough is supplied with two plates of copper, so bent that they could carry a copper

---

*Translation from Thomson's *Annals of Philosophy*, October 1820. Translated from a printed account drawn up in Latin by the author and transmitted by him to the Editor of the *Annals of Philosophy*.

rod, which supports the zinc plate in the water of the next trough. The water of the troughs contained one-sixtieth of its weight of sulphuric acid, and an equal quantity of nitric acid. The portion of each zinc plate sunk in the water is a square whose side is about 10 in. in length. A smaller apparatus will answer provided it be strong enough to heat a metallic wire red hot.

The opposite ends of the galvanic battery were joined by a metallic wire, which, for shortness sake, we shall call the *uniting conductor*, or the *uniting wire*. To the effect which takes place in this conductor and in the surrounding space, we shall give the name of the *conflict of electricity*.

Let the straight part of this wire be placed horizontally above the magnetic needle, properly suspended, and parallel to it. If necessary, the uniting wire is bent so as to assume a proper position for the experiment. Things being in this state, the needle will be moved, and the end of it next the negative side of the battery will go westward.

If the distance of the uniting wire does not exceed three-quarters of an inch from the needle, the declination of the needle makes an angle of about 45°. If the distance is increased, the angle diminishes proportionally. The declination likewise varies with the power of the battery.

The uniting wire may change its place, either towards the east or west, provided it continue parallel to the needle, without any other change of the effect than in respect to its quantity. Hence the effect cannot be ascribed to attraction; for the same pole of the magnetic needle, which approaches the uniting wire, while placed on its east side, ought to recede from it when on the west side, if these declinations depended on attractions and repulsions. The uniting conductor may consist of several wires, or metallic ribbons, connected together. The nature of the metal does not alter the effect, but merely the quantity. Wires of platinum, gold, silver, brass, iron, ribbons of lead and tin, a mass of mercury, were employed with equal success. The conductor does not lose its effect, though interrupted by water, unless the interruption amounts to several inches in length.

The effect of the uniting wire passes to the needle through glass, metals, wood, water, resin, stoneware, stones; for it is not taken away by interposing plates of glass, metal or wood. Even glass, metal, and wood, interposed at once, do not destroy, and indeed scarcely diminish the effect. The disc of the electrophorus, plates of porphyry, a stoneware vessel, even filled with water, were interposed with the same result. We found the effects unchanged when the needle was included in a brass box filled with water. It is needless to observe that the transmission of effects through all these matters has never before been observed in electricity and galvanism. The effects, therefore, which take place in the conflict of electricity are very different from the effects of either of the electricities.

If the uniting wire be placed in a horizontal plane under the magnetic needle, all the effects are the same as when it is above the needle, only they are in an opposite direction; for the pole of the magnetic needle next the negative end of the battery declines to the east.

That these facts may be the more easily retained, we may use this formula—the pole *above* which the *negative* electricity enters is turned to the *west*; *under* which, to the *east*.

If the uniting wire is so turned in a horizontal plane as to form a gradually increasing angle with the magnetic meridian, the declination of the needle *increases*, if the motion of the wire is towards the place of the disturbed needle; but it *diminishes* if the wire moves further from that place.

When the uniting wire is situated in the same horizontal plane in which the needle moves by means of the counterpoise, and parallel to it, no declination is produced either to the east or west; but an *inclination* takes place, so that the pole, next which the negative electricity enters the wire, is *depressed* when the wire is situated on the *west* side, and *elevated* when situated on the *east* side.

If the uniting wire be placed perpendicularly to the plane of the magnetic meridian, whether above or below it, the needle remains at rest, unless it be very near the pole; in that case the pole is *elevated* when the entrance is from the *west* side of the wire, and *depressed*, when from the *east* side.

When the uniting wire is placed perpendicularly opposite to the pole of the magnetic needle, and the upper extremity of the wire receives the negative electricity, the pole is moved towards the east; but when the wire is opposite to a point between the pole and the middle of the needle, the pole is moved towards the west. When the upper end of the wire receives positive electricity, the phenomena are reversed.

If the uniting wire is bent so as to form two legs parallel to each other, it repels or attracts the magnetic poles according to the different conditions of the case. Suppose the wire placed opposite to either pole of the needle, so that the plane of the parallel legs is perpendicular to the magnetic meridian, and let the eastern leg be united with the negative end, the western leg with the positive end of the battery: in that case the nearest pole will be repelled either to the east or west according to the position of the plane of the legs. The eastmost leg being united with the positive, and the westmost with the negative side of the battery, the nearest pole will be attracted. When the plane of the legs is placed perpendicular to the place between the pole and the middle of the needle, the same effects recur, but reversed.

A brass needle, suspended like a magnetic needle, is not moved by the effect of the uniting wire. Likewise needles of glass and of gum lac remain unacted on.

We may now make a few observations towards explaining these phenomena.

The electric conflict acts only on the magnetic particles of matter. All non-magnetic bodies appear penetrable by the electric conflict, while magnetic bodies, or rather their magnetic particles, resist the passage of this conflict. Hence they can be moved by the impetus of the contending powers.

It is sufficiently evident from the preceding facts that the electric conflict is not confined to the conductor, but dispersed pretty widely in the circumjacent space.

From the preceding facts we may likewise infer that this conflict performs circles; for without this condition it seems impossible that the one part of the uniting wire, when placed below the magnetic pole, should drive it towards the east, and when placed above it towards the west; for it is the nature of a circle that the motions in opposite parts should have an opposite direction. Besides, a motion in circles, joined with a progressive motion, according to the length of the conductor, ought to form a conchoidal or spiral line; but this, unless I am mistaken, contributes nothing to explain the phenomena hitherto observed.

All the effects on the north pole* above-mentioned are easily understood by supposing that negative electricity moves in a spiral line bent towards the right, and propels the north pole, but does not act on the south pole. The effects on the south pole* are explained in a similar manner, if we ascribe to positive electricity a contrary motion and power of acting on the south pole, but not upon the north. The agreement of this law with nature will be better seen by a repetition of the experiments than by a long explanation. The mode of judging of the experiments will be much facilitated if the course of the electricities in the uniting wire be pointed out by marks or figures.

I shall merely add to the above that I have demonstrated in a book published 5 years ago that heat and light consist of the conflict of the electricities. From the observations now stated, we may conclude that a circular motion likewise occurs in these effects. This I think will contribute very much to illustrate the phenomena to which the appellation of polarization of light has been given.

Copenhagen, 21 July, 1820   JOHN CHRISTIAN OERSTED

Note by R.A.R.T. Oersted's expressions are "Omnes in polum septentrionalem" and "Effectus in polum meridionalem". If by "a spiral line bent towards the right" he means a right-handed screw, then he must be using septentrionalem in the same sense that boreal was used at the time, namely to indicate a south-seeking pole. The term north pole in the translation would therefore mean a south-seeking pole—i.e. one homologous with the earth's north pole.

2

B   Biot and Savart

I.   NOTE ON THE MAGNETISM OF VOLTA'S BATTERY*

At the Académie des Sciences in its session of 30 October 1820, MM. Biot and Savart presented a dissertation on the determination by precise measurement of the physical laws governing the action on magnetized bodies, of metal wires when in contact with the two poles of a voltaic apparatus. For the experiments, tempered steel rectangular plates or cylindrical wires, magnetized by the method of double contact, were suspended from cocoon threads, and their oscillation time and equilibrium position were observed when suspended at various distances in different directions relative to the metal wire connecting the two poles of the battery. Sometimes the action of terrestrial magnetism was combined with that of the wire and other times it was compensated and destroyed by the opposing action of an artificial magnet placed at some distance away. A trough type of apparatus was used with ten pairs of troughs 1 dm² in surface area. Alternative observations were made which corrected any progressive variations that might have occurred. Time was measured by an excellent half-second double-stop Bréguet chronometer.

By these procedures MM. Biot and Savart arrived at the following result which rigorously represents the action experienced by a molecule of austral or boreal magnetism when placed at some distance from a fine and indefinite cylindrical wire which is made magnetic by voltaic current. Drawing a perpendicular to the axis of the wire from the point where the magnetic molecule resides, the force influencing the molecule is perpendicular to this line and to the axis of the wire. Its intensity is inversely proportional to the

---

*Ann. Chim. Phys. 15, 222–3 (Translated by O. M. BLUNN).

distance. The nature of the action is the same as that of a magnetized needle which is placed on the contour of a wire in a certain constant direction in relation to the direction of the current; thus the molecule of boreal magnetism and the molecule of austral magnetism are influenced in opposite directions, though always in the same straight line, as determined by the foregoing construction.

By this law one can predict and calculate all the motions imparted to magnetized needles by a connecting wire, whatever the relative direction of the wire. The direction of the type of magnetization which can be imparted to steel or iron wires when the action is sustained in a given direction in relation to its length can also be deduced from the ordinary laws of magnetic action.

II. EXTRACTS FROM *PRECIS ELEMENTAIRE DE PHYSIQUE*†

Magnetization of Metals by Electricity in Motion

(i)*

I shall completely abandon the historical order of the discoveries and indicate quickly what has already been done to complete the analysis of electromagnetic forces which M. Oersted so skilfully began.

The first thing which had to be discovered was the law governing the decrease of the force of a conducting wire with increasing distance from its axis. This was the object of the work which I undertook with M. Savart, whose ingenious discoveries in acoustics I have already reported. We took a magnetized steel needle in the form of a very short parallelogram, such as AB in Fig. 41, and to make it perfectly mobile, we suspended it in the horizontal position in a glass cage on a single silkworm thread. To make it quite free to obey the force of the connecting wire, we eliminated the force of terrestrial magnetism by placing a bar magnet $A'B'$ at a distance and in a direction to balance this force exactly. Such compensation is always possible because, whatever the cause of the magnetism which

---

†*Précis Elémentaire de Physique*, Vol. II, 3rd edition, 1824 (*Translated by* O. M. BLUNN).
*Pages 707–723.

the earth exerts on magnetized bodies, and, equally, whatever the mode of distribution of the forces which are produced by it, it is at least certain that the action of these forces makes itself felt everywhere in a resultant force which influences the integral particles of

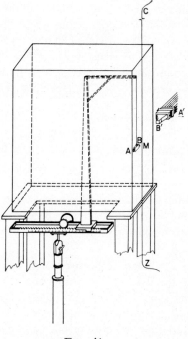

Fig. 41.

the bodies in a certain determinate direction, whence it follows that the influence, always very faint, can be offset by the action of a magnet placed in such a way that it produces on the body under consideration an equal but directly opposite effect. To strike this balance in our hemisphere, where the magnetic resultant of the earth acts as boreal force, it is first necessary, if the needle in question is horizontal, to allow it to turn freely in the magnetic meridian and to observe with a chronometer the number of oscillations under the sole influence of the terrestrial force in a set period of time. Then, taking a stable bar magnet of the greatest possible length and energy, it is placed horizontally at the same height as the needle and on the continuation of its magnetic meridian, whether to north or south, in order to turn it in the opposite direction to the action of the earth, that is to say, so that its boreal pole points north and its austral pole south. If at first the bar is far from the needle, the resultant of the forces which it exerts is very faint, or even imperceptible; this can be checked by making the needle oscillate, because the rate of oscillation will be almost the same as for terrestrial influence alone; but by bringing the bar closer, little by little, the oscillations of the needle become slower, and gradually a position is reached where the oscillation is such that the total resultant still influencing it is altogether negligible. This can readily be seen from the oscillation, at least when the energy of the bar is very great compared with the length of the needle, as recommended. In this condition each pole of the needle is noticeably acted upon in the same way by the bar in parallel directions wherever the oscillatory motion may take it. Now this parallelism of direction takes place equally for the terrestrial force, and in an infinitely more rigorous way. The oscillatory motion due to the difference between these two actions is therefore like that which would be obtained by the influence of a single very faint directing force acting always in apparently parallel directions; this is what makes the squares of the oscillation times inversely proportional to the intensities of the force when the oscillations are very low in amplitude. The residue of the force which persists in any position that one might put the bar, is thus known and the position where the oscillation becomes slow enough for the terrestrial force to be regarded as zero is selected. If the distance from the bar is still very large compared with the dimensions of the needle, as we have assumed, the same compensation will still persist, noticeably to the same extent, in any position of the needle about its centre due to the effort of other influences which in no way change its magnetic state: the needle may then be regarded as perfectly free, as though the earth did not exist, or as if one were transported with it into outer space. We shall see later how any imperfection in this overall neutralization can be corrected rigorously.

Such was the state of equilibrium to which we brought the small magnetized needle which we used in the experiment. When we had satisfied ourselves on this, we passed current through the cylindrical copper connecting wire ZC. This wire had been placed vertically in front of the needle at a sufficient distance away. It was long enough for its extremities to be bent back and connected to the poles of the battery and still only exert such a feeble effect on the needle that it could confidently be ignored. This arrangement represented the effect of an infinite vertical wire acting on a free and horizontal magnetized needle. As soon as the current began to flow, the needle turned transversally to the axis of the wire, in conformity with the rotary behaviour indicated by M. Oersted; it then began to oscillate about this direction, just as the stem of a pendulum will oscillate about the vertical due to the effect of the weight; finally, it settled in this direction when the excursions had been stopped by the resistance of the air. The progressive gradual approach of the needle to this definite position was sufficient to indicate that the state of equilibrium was of the type which is called stable; in fact, if it was moved only ever such a little and then left free to swing, it returned to the same place after its oscillations. To determine the nature of the resultant force which returned it, we set the needle slightly in motion and, using a Bréguet half-second chronometer, we counted the time required to complete a certain number of oscillations, twenty for example, and then counted on in sets of twenty for as long as the excursions were large enough to be observable. We satisfied ourselves by these tests that their duration was noticeably independent of their amplitude within the limits under consideration. Now, when a solid body of prismatic shape, such as our needle, is free to turn about the axis passing through its centre and oscillates about a certain equilibrium position, if it behaves with regular periodicity in the oscillations which return it, it may be inferred that the force which makes it turn is exactly, or almost exactly, proportional in all its successive positions to the angle through which it is moved from the direction; hence the isochronism (regular periodicity) of the motions, since it is constantly called to its point of rest with energy which is noticeably proportional to the angle which remains for it to describe in order to arrive there. The motion of a solid body at these low amplitudes may be rigorously likened to the motion of a simple pendulum which oscillates about an equilibrium position due to gravity. Now the oscillations of such a pendulum, if of constant length, vary in duration according to the intensity of the weight influencing it, and this intensity is reciprocally proportional to the squares of the times taken by the pendulum to complete a number of very low amplitude oscillations. Likewise, if the squares of the times for different distances between the wire and the needle are compared, assuming that the condition of isochronism is fulfilled, the ratios of the component forces exerted by the wire parallel to the direction of equilibrium about which the needle oscillates become known. These ratios, and the possibility of equilibrium, are therefore all conditions which the total force of the wire must satisfy; consequently, the absolute law governing this force can be discovered for these conditions to hold.

But to discover the nature of the force in this way, it is first necessary to analyse the way in which the force can influence the needle and determine its motion. I have said that we took care to use a very short needle. We had so magnetized it that it was exempt from consequent points; its free quantities of austral and boreal magnetism could therefore be regarded as concentrated at two points or poles near to each extremity of the needle at equal distances between them; if the needle had been a single very fine cylindrical wire, these poles would, as is known, have been separated a sixth of the total length. Now, since these two poles are opposite in nature, the influence of the connecting wire, whatever it may be, must also be opposite on each of them, that is to say, if the equal quantities of austral and boreal magnetism which reside there are placed at one and the same point, and the influence of the wire induces one of them in a certain direction, it should induce the other in the absolutely opposite direction, and with equal energy. This condition of opposition is implied from the fact that a conducting wire only moves non-magnetized needles which are placed near to them after having separated their natural magnetisms; they thus remain without power when made to act on tempered steel needles, or needles made from other very hard alloys in which the natural magnetisms have not been separated beforehand, and the effort required for separation is too great. Such impotence would not occur if the molecules of the combined magnetisms are induced in different directions, but not exactly opposite to each other; the resultant of

the various efforts exerted by the wire on the external bodies, in the
state of combination of the magnetisms, is not then zero; and in
consequence, needles made from magnetic metals would commence
to move under the influence of connecting wires without being even
temporarily magnetized, which is contrary to the phenomena. The
indifference of these needles while their natural magnetisms are
unseparated likewise proves the equality of the two magnetisms;
for, without this equality, even needles not actually magnetized, but
formed of magnetic substances, would, in the presence of these wires,
acquire absolute translatory motion in space.

Using these principles for our needle in the position of equilibrium
assigned to it by the influence of the connecting wire, we draw
through its magnetic axis a horizontal plane which is therefore
perpendicular to the wire. This plane contains all the forces by which
equilibrium is determined; for it contains the two poles of the needle
where the two quantities of free magnetism reside which undergo

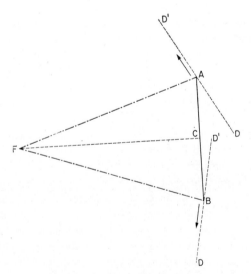

FIG. 42.

the action of the wire, and it also contains the resultant of this action on the poles, since the wire is regarded as indefinite and the effects which it produces are the same, wherever the needle is presented over its length, the two parts of the wire on either side of the horizontal plane through the centre of the needle necessarily exert the same action upon it; hence their common resultant for each pole must be in the same direction as the plane itself. These results are illustrated in Fig. 42, where AB is the magnetic axis of the needle, A and B are its two poles, C is its centre, and F is the projection of the wire in the horizontal plane, or its intersection with this plane. Now, whatever the nature of the action exerted by the wire on pole B, this action takes a direction in the plane FBA. Suppose that it is BD, and the quantity of free boreal magnetism at B is induced along BD due to this force. Obviously, if the point B were occupied by a like quantity of austral magnetism, this would be induced in the directly opposite direction with equal force, that is to say, it would move along BD', which is a continuation of BD. The direction and intensity of the action exerted by the wire on the austral pole A of the needle is not obtainable directly; the pole A has the same quantity of free magnetism as B; moreover, it is a distance FA equal to FB from the wire in the equilibrium position where the needle settles. Now it can readily be seen that the actions of the wire are the same at the same distance from the centre before whichever point of the contour the needle is placed; for if the wire is made to turn on itself without changing either its longitudinal direction or its distance from the needle, the oscillatory motion of the needle and its relative equilibrium position are not affected. Consequently, to find the action exerted at A on a molecule of austral or boreal magnetism, the same construction has only to be repeated as at the point B, that is to say, the line DAD' is drawn at an angle to FA equal to the angle FBD; taking two equal and opposite lengths AD, AD' on this line, from point A, the former represents the force which would influence a quantity of boreal magnetism at A, whilst the other represents the force which would influence an equal quantity of austral magnetism situated at the same point. This case is precisely the same as when the needle is in its equilibrium position. Its mass is in fact influenced by two resultants: the one force BD applied to pole B draws it from B towards

D; the other force AD', equal to the foregoing, but applied at pole A, draws it from A towards D' with the same intensity. Now it is evident from the simplest of static considerations that since the two forces BD, AD' are equal, applied to two lever arms CA, CB of equal length, and since they tend to make the needle point in opposite directions, they cannot leave it in equilibrium in the position BCA which is perpendicular to CF unless their directions BD, AD' are equally inclined to the needle, that is to say, unless the angles DBC, $D'AC$ are equal; this requires that DBF be equal to $D'AF$. But $D'AF$ is identically equal by construction to $D'BF$ since the system of lines FA and DAD' is nothing else than the system of lines FB and DBD' transported from B to A. Consequently, in the observed state of equilibrium the two angles DBF, $D'BF$ are equal to each other; hence it follows that they are both right angles since they are adjacent on one and the same straight line. *Thus, when an indefinite connecting wire, animated by voltaic current, acts on an element of austral or boreal magnetism situated at a certain distance FA or FB from its centre, the resultant of the actions which it exerts is perpendicular to the shortest distance between the molecule and the wire.*

We do not have to examine here how the various cross-sections of the wire contribute to form this resultant, nor do we have to consider how the infinitesimal molecules of each section can produce resultants transversal to the wire. These factors necessarily depend on the nature of the forces developed by the current in each integral molecule composing the mass. Knowledge of these molecular forces would no doubt be very useful, and perhaps highly desirable for that reason, but to establish the reality or direction of their resultant such knowledge is not at all necessary. As we have seen, observation itself of the component results is sufficiently rigorous.

It is now necessary to determine the absolute direction of this resultant when acting upon each type of magnetism, that is to say whether it tends to entrain it to the right or to the left of the lines FA and FB. It is also necessary to determine the law governing its variation with distance. This is precisely what Savart and I did in the two series of experiments which I am about to report.

The first series of experiments was carried out with the above apparatus; it was sufficient to make the support of the connecting wire mobile so that it could be presented to the needle at different distances as desired; this was arranged by fixing a horizontal section at the base of this support in the direction of the suspension thread of the oscillating needle, the support being free to move along it. If the horizontal distance from the thread or centre of the needle to the connecting wire is then measured directly for some position of the wire, it is clear that all the other distances can be obtained by adding to, or subtracting from, this distance all the amounts which the support of the connecting wire will have travelled along the horizontal section. To discover the action of the wire separately without the intervention of any external force, the action on the needle of terrestrial magnetism is neutralized by a powerful magnet appropriately placed at a great distance away as explained above. Having made these arrangements, the position of the connecting wire is adjusted successively to different distances, but with a sufficient distance for the duration of the small oscillations of the needle under its influence to be perceptibly isochronal; the number of seconds and half-seconds taken by the needle to perform a certain constant number of oscillations, say ten, is then counted with every possible care in each succeeding position; for greater exactitude, larger numbers of oscillations may be counted. Since the isochronism of the oscillations allows the motion to be regarded as if it were produced by a force parallel to the directions of the equilibrium position in which the needle settles, it follows that this case is precisely that of a pendulum which is made to oscillate successively in different latitudes about the vertical under the influence of weights of different intensity. Hence, for both needle and pendulum, the energy of the component force, thus directed, is reciprocal to the squares of the oscillation times. Putting F for the particular ideal force which would make the needle oscillate exactly once in ten seconds, in another position of the wire where the same number of oscillations is performed in a different number of seconds N, the force is no longer F, but $100F/N^2$. Thus, it only remains to compare these forces at corresponding distances and discover their relationship.

But some precautions must be taken in the experiments to make consecutive results exactly comparable; otherwise grave errors may occur. First, the flow of current must not be continuous to avoid running down the battery and progressively weakening its action so that it itself is no longer comparable. The apparatus must be fitted with troughs where the connectors may only be immersed during the observations; so as to be able to transmit current at will through different wires, e.g. ZMC, $Z'M'C$ (Fig. 43) or through the same wire several times, without risk of differences due to imperfect contact, the ends Z, C, Z', C' of these wires are immersed and fixed in separate mercury-filled glass jars; then, to transmit the current through the wires in any desired direction, it is sufficient to plunge the ends of the conducting wires from the two extremities, zinc or copper, of the battery momentarily into the jars; these must be soldered to the plates which terminate it, not attached. Also, if the needle does not enter into oscillation promptly enough, it is necessary to avoid exciting it by contact with a solid body, which would always impart to it agitations that would shift it off-centre and would take long to subside as the suspension thread was excessively mobile; it is necessary to displace it by presenting a piece of soft iron from a distance to one of its poles for an instant, which, being magnetized under its influence, attracts the needle. It is also necessary to define the oscillation time exactly by not counting the oscillations from the instant at which they begin or finish; at the reversal of motion the needle remains momentarily stationary and it is therefore impossible to fix on the precise instant corresponding to the start or end of oscillation. This indeterminacy can be avoided by stretching two very fine silk threads vertically in the continuation of the needle's equilibrium position and counting the oscillations from the instant when the needle passes the thread to that when it comes by again, because its movement being most rapid at these times, the change from one oscillation to the next can be determined more surely than at any other point of its amplitude. Nevertheless, apart from this arrangement, it is also necessary to take the extra simple precaution of always counting an even number of oscillations, and not an odd number; for if the sight thread is not placed exactly at the mid-point of the total oscillation amplitude, and it is impossible to eliminate all error in this respect, the time taken by the needle to go by the thread and return will be under- or over-stated; but the time taken will be as short on one side as it is too long on the other; the error is therefore compensated exactly in a pair of consecutive oscillations and it is therefore compensated likewise in any even number of oscillations. Finally, for the results of successive experiments to be perfectly comparable, despite the progressive weakening of the battery, it is necessary to employ the method of alternatives, like that used in experiments with electricity. That is to say, should it be desired, for instance, to compare the actions of a particular wire at different distances, the needle is first made to oscillate at a certain distance D from this wire, then at another distance D', and then again at the first distance D, and so on thereafter, coming back each time to the first distance and taking care to make the partial experiments as nearly equal as possible in duration. Then, taking the average of the first and third observations, the result for the distance D is comparable with the observation made at the distance D', and so on for all the other observations. All these precautions were taken by M. Savart and me for our measurements. The prismatic needle which we used was 20 mm long, 10 mm wide, and 1 mm thick; this gives 10 mm for the length of the oscillating arm on each side of the centre. But, according to the laws governing the distribution of free magnetism in magnetized bodies, each portion of these arms did not have to be equally charged. If our needle had been a cylindrical wire, its two poles, that is to say, the magnetic centres of gravity of its two arms, would have been placed at two-thirds of each of them, that is to say, at 6·6 mm from the centre; thus the actions exerted by the connecting wire on the two poles would have had to be considered as applied solely to the extremity of this reduced length; but it may be assumed that the actual distance between the poles of our needle was less than this limit, whether due to its prismatic form, or to the fact that it had been left for a long period of time to its own magnetic reaction, which must have fundamentally weakened the excess of charge of the extremities where magnetism is always most developed. At least, this is what we had to infer from the experiments themselves, seeing that the oscillations of the needle conserved their isochronism and independence of amplitude at distances from the wire which were not such that they could be deemed very large compared with the half-length. In any event our findings are given in the Table I. On the right I have placed the ratios of the resulting forces according to the squares of the oscillations, using the results obtained at the distance of 30 mm for the reference.

TABLE I.

Order of the observations	Distances from the connecting wire to the centre of the needle (mm)	Duration of ten oscillations (sec)	Ratio of the observed forces, compared with the force at 30 mm
1	30	42·25	
2	40	48·85	3/4 (1—0·008508)
3	30	42·00	
4	20	33·50	3/2 (1 +0·023090)
5	30	41·00	
6	50	54·75	3/5 (1—0·036673)
7	30	42·25	
8	60	56·75	2/6 (1 +0·095460)
9	30	41·75	
10	120	89·00	3/12 (1—0·103892)
11	30	42·50	
12	15	30·00	2/1 (1 +0·067010)
13	30	43·15	

The numbers in the last column show that the ratios of the observed forces are almost exactly inverse to the ratios of the distances to the connecting wire. To discover whether this simple law is valid within the limits of error of these observations, it need only be used to calculate hypothetically the duration of the oscillations in each experiment and compare it with those at 30 mm; for if N is the number which refers to the distance 30 mm and N' is the number required for the distance D', we have

$$\frac{N'^2}{N^2} = \frac{D'}{D};$$

hence
$$N' = N\sqrt{(D'/D)}.$$

Table II is calculated in this way.

TABLE II.

Order of the observations	Distances to the wire	Duration of ten oscillations		Difference (sec)
		Calculated (sec)	Observed (sec)	
2	40	48·62	48·85	—0·23
4	20	33·88	33·50	+0·38
6	50	53·74	54·75	—1·01
8	60	59·40	56·75	+2·65
10	120	84·25	89·00	—4·75
12	15	30·99	30·00	+0·99

The errors are alternately positive and negative, but without any regular law; they are considerable at large distances, but this is to be expected, for we had not dreamt of making the apparatus which carries the connecting wire mobile in the first experiments; instead, the needle was made to move and we adjusted the compensation in each experiment by moving the magnets. The feeble action which persisted after the force of earth had been compensated in the best possible way, must still have exerted relatively greater influence at greater distances where the action of the connecting wire is weakest compared with nearby where it is strongest. Explaining the discrepancies in this way, they can be safely ignored and we can adopt the foregoing law as being as exact as the observations themselves; but, still remaining within the limits of error, a more probably physical interpretation can be placed on the proportionality which this law discloses if it is assumed to apply, not to the distance from the centre of the needle to the wire, but to the distance between the wire and the two poles of the needle. These distances must necessarily be different; yet if the magnetic centres of gravity of the two arms of our prismatic needle are closer to the centre than those of cylindrical wires, as seems highly probable, the distance from the wire to the centre and to the poles can be so nearly the same that the slight difference can be ignored in calculations. Interpreted in this way, the law implies that *the total action of a connecting wire on some magnetic element, whether austral or boreal, is reciprocal to the rectilinear distance between this element and the wire.*

To justify this result completely, I performed another series of the same experiments still using a parallelogram-shaped prismatic needle, but a much shorter one; its length was only 10 mm, its width and thickness being 5 mm and 0·5 mm respectively. The approximation was obviously more rigorous in having the magnetic centre of gravity of each arm nearer to the centre of the needle than before. I suspended this needle again by a cocoon thread and made it oscillate at different distances from the vertical connecting wire. But instead of neutralizing the action of earth by the opposition of a distant magnet, I merely enfeebled it, taking care to approach the magnet in a direction such that the needle did not deviate from its natural magnetic meridian, and I placed the apparatus which carried the vertical connecting wire so that this wire was exactly contained in the plane drawn through the centre of the needle perpendicularly to the same meridian. With this arrangement the force of the connecting wire in no way tended to deflect the needle from its magnetic meridian; it only increased or decreased the resultant which held it there. Before allowing the wire to influence it, I first measured the resultant by counting the number of seconds required for a set number of oscillations. Let N be this number. Repeating the same observation with the wire influencing the needle I measured the sum of all the forces which act upon it in the new arrangement. Putting N' for this new number, according to the laws of oscillatory motion, the first system of forces is represented by K/N^2, and the other by K/N'^2, where K is a constant which depends on the dimensions of the needle; the difference

$$\frac{K}{N'^2} - \frac{K}{N^2} \quad \text{or} \quad \frac{K(N-N')\ (N+N')}{N^2N'^2},$$

therefore measures the force exerted by the connecting wire alone.

Here (Table III) the method of alternatives has been used again in order to avoid the effect of any variation in the energy of the battery during the experiments. This accounts for the slight difference between the first and third results, even though the distance from the wire was the same. The closeness of the results indicates that the battery was practically constant. I had in fact made it from a single voltaic element of very large dimension which could be plunged as

TABLE III.

Order of the observa-tions	Distances from connecting wire to centre of needle (mm)	Duration of forty oscillations		Action of the wire alone	Ratio of the actions of the wire at the distances 32·9 mm and 62·9 mm
		Without influence of the wire N (sec)	With influence of the wire N' (sec)		
1	32·9	67·5	101·0	1·21449	
2	62·9	66·0	78·5	0·67290	$\dfrac{32\cdot9}{62\cdot9}(1+0\cdot03781)$
3	32·9	66·0	98·5	1·26499	

desired into a large circular tank; I left it immersed there throughout the three experiments because observation of ignition phenomena in voltaic wires, as well as of oscillation, shows that the energy of the apparatus is always greater in the first few moments after immersion than shortly afterwards when it has become steady. This more stable state is obviously to be chosen for comparison experiments. The effect of small variations which may occur later is completely eliminated by taking the average of the first and last results at the same distance; it is this average which must be compared with the intermediate result. This comparison is given in the last column of Table III where it can be seen that the difference between the ratio of the forces and the inverse ratio of the distances is a very small fraction. The physical interpretation which we have placed on this law is thus confirmed in reference to magnetic elements themselves.

(ii)†

The action of an indefinite and rectilinear connecting wire on a magnetic element, such as that obtained in the foregoing experi-

---

†Pages 740–746.

ments, is still nothing but a composite result; for by imagining the length of wire to be divided into infinitely many fine sections of very low height, it is seen that each section must act on the needle with a different energy according to its distance and direction. Now these elementary forces are just the simple result which it is specially important to know, for the total force exerted by the wire is nothing other than the arithmetic sum of these effects. However, calculation is sufficient to rise from the resultant to the simple action. This is what Laplace did. From our observations he deduced mathematically the law of the force exerted individually by each section of the wire on each magnetic molecule presented to it. This force, like the total action, is perpendicular to the plane drawn through the longitudinal element of wire in the shortest distance between this element and the magnetic molecule which is influenced. The intensity of the force, as in other magnetic actions, is reciprocal to the square of the distance. However, the distance ratio may be modified by a coefficient which depends on the inclination of each distance to the general direction of the wire; that is to say, such a coefficient, whatever its composition, does not prevent the total action of a straight and indefinite wire being reciprocal to the shortest distance to the magnetic element, in accordance with our observations. It was therefore necessary to carry out fresh experiments to find out whether the coefficient did in fact exist, and to discover how it was composed; the simplest and most direct way of doing this was obviously to compare the actions exerted on the same magnetic element by two equal portions of indefinite wires in different directions. Accordingly, I stretched a long copper wire ZMC, Fig. 43 in the vertical plane, bending it at M so that the two arms ZM, MC should be at the same angle to the horizontal. In front of this wire I stretched another piece $Z'M'C'$ of the same metal, the same in diameter and of the same drawing; this piece I set up vertically, being separated from the first piece at MM' only by a strip of very fine paper. I then suspended the small parallelogrammic magnetized needle AB in front of this system so that its longitudinal axis was at the height of the points M, M' exactly, and I observed the oscillations at various distances whilst passing current successively through the bent and straight wires. In such a comparison it is always necessary to use the method of alternatives, that is to say, if the oscillations have been

observed in one position of the needle, with the current flowing
through the oblique wire and then through the straight wire, it is
necessary to observe them a third time with the current again
flowing through the oblique wire, and then to take the average of the

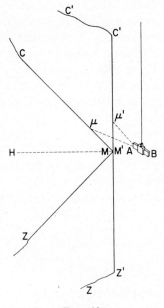

Fig. 43.

results of both observations for otherwise the action of the oblique
wire may not be exactly comparable with that of the straight wire
owing to progressive variation of the voltaic apparatus. But, to be
precise, experiments of this kind require other precautions to be
taken; first, the two wires must be perfectly identical, and this can be
arranged as far as possible by taking both parts from one continuous
piece of wire of the same drawing; secondly, the oblique and straight
wires must be hung exactly in the same vertical plane, and the centre
of the needle should also be in the continuation of this plane in all
its successive positions. Finally, for the line of rest about which
the needle oscillates to remain invariably the same, whichever of

the two wires is made to act upon it, it is essential, if the effects of the actions of the wires are to be comparable at all distances, for the plane of the two wires to be perpendicular to the natural magnetic meridian of the needle and for the terrestrial force which influences it to be enfeebled by an artificial magnet without in any way altering the direction of the needle. As a result the action of the wires on the needle, whether the oblique or vertical, is in fact exerted in this same direction and, in consequence, it only intensifies the force which returns it, which can be seen from the rapidity of the oscillation in other cases which we have considered as well. By faithfully observing every precaution, the results for the actions of each wire are in perfect agreement with each other after taking off the primitive, directing force independent of the wires which has been allowed to persist; and similar agreement is obtained whatever the intensity or direction of this force, provided that it is not so feeble that the situation of the needle is deranged by the action of the wires when the current is transmitted in the opposite direction; for, unless the external magnet which enfeebles the terrestrial directing force is remote from the needle in relation to its dimensions, such a change will always slightly alter the effect on the needle and our inherently delicate results are necessarily affected, wherefore no care is too great to preserve their reliability. I did not realize the indispensable necessity of all these precautions at first. The early tests soon made me see that the action of the oblique wire diminishes with decreasing angle between the two arms, and this seemed to be proportional; such a law was quite compatible with the limits of the phenomenon, for the action must obviously be zero if the angle is zero when the two halves of the wire are bent together with the current flowing in opposite directions; and this action and that of the straight wire must become equal when each arm is at 90° to the horizontal, since they then both form a vertical straight line. But because of imperfections in the experiments, other laws were equally admissible and, for example, the inclination i to the horizontal could have been replaced by the tangent to half of the inclination, that is to say, $\tan \frac{1}{2}i$; then, by putting F for the observed action on the needle of the vertical wire at a certain distance, $F \tan \frac{1}{2}i$ would be the action of the oblique wire animated by the same current; formerly, it would have been represented by $Fi/90°$, a value which can only ever differ from the foregoing in hundredths. To check this alternative it was only necessary to repeat the experiment for one angle, but with extreme rigour. I elected to study the case $i=45°$, where the action of the oblique wire is equal to F tan $22°30'$, or F. $0 \cdot 414214$. Since the coefficient of F was very close to $\frac{1}{2}$, I bent the oblique wire double in such a way, without contact between its parts, that in passing twice before the needle it would exert double the action upon it, thereby facilitating observation through the enlargement. I also took care to change the contact between the two wires and poles of the voltaic apparatus so that each of the two poles should terminate alternately above and below each wire, as indicated by the lettering ZB, CH and CB, ZH. Since the action of the wires on the needle is reversed in these two systems of communication, it is seen that the primitive directing force is increased in the one system and reduced in the other. Suppose, therefore, that N is the number of seconds taken by the needle to complete a constant number of oscillations solely under the influence of the primitive force, and that N' were the analogous number when the action of the wire is additive; it is seen without difficulty that the action of the wire itself must be

$$\frac{K}{N'^2} - \frac{K}{N^2} \quad \text{or} \quad K\frac{(N-N')(N+N')}{N^2 N'^2} \, ,$$

K being a constant depending on the dimensions of the needle; conversely, it becomes

$$\frac{K}{N^2} - \frac{K}{N'^2} \quad \text{or} \quad K\frac{(N'-N)(N+N')}{N^2 N'^2}$$

if the action is subtractive; this permits evaluation in either case. In this way Tables IV and V have been compiled from experiments with two small rectangular needles which I described above.

The value of $2F$ tan $\frac{1}{2}i$ yields $0 \cdot 828427$ for this ratio; in these experiments the difference is slight. But it becomes still smaller when it is considered that with the two arms of the oblique wire each deflected 3 mm to right and left of the vertical wire, their distance to the centre of the needle is not d, as for this wire, but $(d^2+9)^{1/2}$ or $d(1+9/d^2)^{1/2}$, which can be reduced to $d[1+(9/2d^2)]$ owing to the

TABLE IV.

Length of the needle (mm)	Duration of twenty oscillations without the wires N (sec)	Duration of twenty oscillations with the influence of the wires N' (sec)		Ratio of the actions of the double oblique wire to the vertical wire O/V
20	58·375	ZN oblique vertical oblique vertical	80·25 87·75 80·00 87·00	0·841713 0·844503
	58·875	ZB oblique vertical oblique vertical	48·19 47·08 48·68 47·27	0·846444 0·830426
		Average		0·840780

TABLE V.

Length of the needle (mm)	Duration of forty oscillations without the influence of the wires N (sec)	Duration of forty oscillations with the influence of the wires N' (sec)		Ratio of the actions of the double oblique wire to the vertical wire O/V
10	67	ZH vertical oblique vertical oblique	102·00 90·50 100·00 89·50	0·806987 0·827775
		ZB vertical oblique vertical oblique	54·00 56·00 53·75 55·50	0·789370 0·802303
		Average Previous average		0·806309 0·840780
		Final average		0·823694

smallness of the fraction $9/d^2$. Thus, to reduce the observations to the same distance for corresponding observations of both wires, it is necessary to multiply the proximate ratio O/V by the inverse ratio of the distances $1+(9/2d^2)$. Now the value of d in the first experiment was 28·5 mm; in the second it was 33 mm; hence the two direct averages must be increased by the fraction $9/2d^2$, which is $\frac{1}{180}$ for the one, and $\frac{1}{242}$ for the other. This correction is 0·845451 for the first and 0·809641 for the second; the final average 0·827545 coincides almost rigorously with 0·828427, which is the value of $F \tan\frac{1}{2}i$. Doubtless this expression generally represents the total action of an oblique wire bent in two arms at the angle i to each other. Now, considering an infinitely fine section of a similar wire situated at μ, Fig. 43, where μm or R is the distance from the wire to the particle m of boreal or austral magnetism, we know from our previous experiments that the action of this section on the particle is reciprocal to the square of the distance μm multiplied by an unknown function of the angle $m\mu M$ for which we put ω. It therefore only remains to find a form for this function such that a resultant proportional to $(\tan\frac{1}{2}i)/2$ if formed by the total sum of the actions of all the wire sections exerted on m perpendicular to the plane CMZ. This condition is satisfied by taking $\sin\omega$ for the required function; this makes the elementary action of a section proportional to $R^{-2}\sin\omega$; using this experimentally determined expression, knowing the absolute direction of the force to be perpendicular to a plane through each distance in the direction of each longitudinal element of wire, the total resultant of the action exerted by a wire, or some straight or curved, limited or unlimited portion of wire, can be assigned by calculation.*

*Note by R.A.R.T. This account confirms Ampère's criticisms. The proportionality of the force to $\tan\frac{i}{2}$ was not suggested by Biot's experiments. It was only confirmed by them after it had been pointed out that the Biot–Savart law did not follow from the previous result on which it had been based, namely that the force was proportional to i.

3
The Early Papers of Ampère

C   DISSERTATION OF M. AMPÈRE*

I. THE MUTUAL ACTION OF TWO ELECTRIC CURRENTS

1. Electromotive action manifests itself in two types of effects which need to be distinguished by precise definition.

I shall call the first effect *electric tension* and the other *electric current*.

Electric tension occurs when the two bodies between which the action takes place are separated† by a non-conducting body over their entire surface except at those points where tension is established; the other effect occurs when these bodies form part of a circuit of conducting bodies by which contact is made at various points on their surface with the points where the electromotive action is produced‡. In the first case, the effect of the electromotive action is to place the two bodies, or two systems of bodies, between which the action takes place, in two states of tension, the difference between which is constant if the action is constant, for example, when it is

---

Ann. Chim. Phys., vol. 15, 1820, pp. 59–76, 177–208 (*Translated by* O. M. Blunn). Dissertation presented to the Académie Royale des Sciences on 20 October 1820, containing a summary of the readings at the Académie on 18 and 25 September 1820, concerning the effects of electric currents by M. Ampère.

†When this separation is due to simple interruption, it is still a non-conducting body, air, which separates them.

‡This is inclusive of the case when the two bodies, or systems of bodies, between which the electromotive action takes place, is in complete communication with the common reservoir which then forms part of the circuit.

due to contact between two substances of different nature; but the difference would vary with the cause which produces it if it were due to rubbing or pressure.

2. But when the two bodies, or two systems of bodies, between which the electromotive action takes place are in contact via conducting bodies between which the electromotive action is not equal and opposite to the first so as to maintain the state of electric equilibrium and hence the tensions, these tensions vanish, or at least become very small, and characteristic phenomena occur. Since the arrangement of the bodies between which the electromotive action takes place is otherwise the same, the action doubtless continues, and since the mutual attraction of the two electricities, as measured by the difference between the electric tensions which has become zero, or else is considerably diminished, can no longer balance this action, it is generally accepted that it continues to carry the two electricities in two senses as before; a double current thus results, the one positive electricity and the other negative electricity, moving in opposite senses from the points where the electromotive action takes place to meet again in the part of the circuit opposite these points. The currents of which I am speaking accelerate until the electromotive force is balanced by the inertia of the electric fluids and the resistance of even the best conductors, whereupon they progress indefinitely at a constant speed so long as the force retains the same intensity; but they cease instantly whenever the circuit is interrupted. For the sake of simplicity I shall call this state of the electricity in a series of electromotive and conducting bodies *electric current*; and since I shall continually have to speak of the two opposite senses in which the two electricities move, I shall invariably imply *positive electricity* by the words *sense of the electric current* to avoid unnecessary repetition; thus, for example, for a battery, the phrase *direction of the electric current in the battery* signifies the direction from the extremity where the hydrogen is disengaged in decomposition of the water to that where the oxygen is obtained; the phrase *direction of the electric current in the conductor which establishes communication between the two extremities of the battery* signifies the opposite direction from the extremity where the oxygen is produced to that where the hydrogen develops. To cover these two cases by a single definition, it may be said that what is called the

direction of the electric current is the direction of the hydrogen and the bases of salts when the water or saline substance of a circuit is decomposed by current, whether these substances form part of the conductor in a battery, or whether they are interposed between the pairs of which the battery is composed.

From the learned researches of MM. Gay-Lussac and Thenard into this apparatus, a fruitful source of great discoveries in almost every branch of physical science, the decomposition of water, salts, etc., is in no way due to the difference in tension between the two extremities of the battery, but solely to what I have called *the electric current*, since the decomposition is practically zero in plunging the two conducting wires into pure water; whereas, without in any way altering the rest of the apparatus, if an acid or saline solution is mixed with one of these substances it conducts electricity well.

Now it is obvious that the electric tension of the extremities of the wires immersed in the liquid could not have been increased in this second case; the tension can only decrease according as this liquid becomes a better conductor; what produces the increase in this case is the electric current; it is therefore solely due to it that the decomposition of the water and of the salts, occurs. It may readily be verified that it is also only the current that acts on the magnetized needle in the experiments of M. Oersted. For this it is sufficient to place a magnetized needle on a horizontal battery situated roughly in the direction of the magnetic meridian; so long as its terminals are not in communication, the needle conserves its ordinary direction. But if a metal wire is attached to one terminal and the other is brought into contact with the extremity of the battery, the needle suddenly changes direction and it remains in its new position so long as contact is made and the battery conserves its energy; it is only to the extent that energy is lost that the needle reverts to its ordinary direction; whereas if the current is made to cease by interrupting the communication, it returns instantly. However, it is this same connection which causes the electric tensions to cease or to decrease considerably; it cannot therefore be these tensions, but the current alone, which influences the direction of the magnetized needle. When pure water forms part of the circuit, and the decomposition is hardly perceptible, a magnetized needle placed above or below another portion of the circuit is deflected just as slightly;

when nitric acid is mixed with the water, without otherwise altering the apparatus in any way, the deflection is increased at the same time as the decomposition of the water is made more rapid.

3. The ordinary electrometer indicates the presence of tension and the intensity of this tension; there used to be no instrument for making known the presence of electric current in a battery or conductor and which would indicate its energy and direction. Such an instrument does exist today; it is sufficient to place the battery, or some portion of the conductor, roughly in the horizontal position in the direction of the magnetic meridian, and to place an apparatus similar to a compass (the only difference being the use to which it is put) on the battery or well above or below the portion of conductor: as long as the circuit is interrupted, the magnetized needle remains in its ordinary position; but it deviates away from it as soon as the current is established, and more so the greater its energy, and the direction can be told if the observer imagines himself to be placed in the direction of the current so that the current flows upwards from his feet to his head when facing the needle, for it is constantly to his left that the action of the current deflects the extremity which is pointing to the north, what I call the *austral pole of the magnetized needle* because it is the pole which is homologous to the south pole of the earth. This is what I express more concisely in saying that the austral pole of the magnet is carried to the left of the current acting on the needle. To distinguish this device from the ordinary electrometer, I think that it ought to be given the name *galvanometer* and it is appropriate to use it in all experiments on electric currents, as one habitually uses an electrometer with electric machines, so as to see if at each instant the current is there and find out its energy.

The first use to which I put this device was to check that the current which exists in the battery between the negative and positive extremities had the same influence on a magnetized needle as the current in a conductor from the positive extremity to the negative.

It was desirable to have for this two magnetized needles, one placed on the battery and the other above or below the conductor; it is seen that the austral pole of the needle is carried to the left of the current near to which it is placed; thus, when the second needle is above the conductor, the needle is carried to the side opposite to that towards which the needle on the battery tends, since the currents are in opposite directions in these two portions of the circuit; the two needles are, on the contrary, carried to the same side, remaining roughly parallel to each other when one is above the battery and the other below the conductor†. As soon as the circuit is interrupted, they immediately revert, in both cases, to their ordinary position.

4. Such are the differences which were known to exist between the effects produced by electricity in its two states which I have just described, the one being, if not a state of rest, at least one of slow motion due solely to the difficulty of isolating bodies in which electric tension occurs, the other being the double flow of positive and negative electricity along a continuous circuit of conducting bodies. In the conventional theory of electricity the two fluids of which it is thought to be constituted, are conceived to be perpetually separated in a part of the circuit and to be carried rapidly in contrary senses into another part of the circuit where they are continually re-uniting. Though such electric current may be produced by arranging a conventional machine so as to develop the two electricities with a conductor to join the two parts of the apparatus where they are produced, the current can only be obtained in large quantities by a voltaic battery, unless very large machines are used, because the quantity of electricity produced by a friction machine is constant throughout a given period, whatever the conduction capability of the rest of the circuit, whereas that which a battery circulates in a similar period increases indefinitely according as the two extremities are connected by a better conductor.

But other more remarkable differences also exist between the two states of electricity. These I have discovered by joining the extremities of two voltaic batteries with two straight parts of two conducting wires in parallel, the one fixed, but the other, suspended from points and made highly mobile by a counter-weight, free to move parallel towards it or away from it. I observed that by passing current through both parts at the same time, they were mutually attracted

---

†For this experiment to leave no doubt as to the action of the current in the battery, it is better to use a trough battery with zinc and copper plates soldered together over the entire interface, and not just simply over a branch of metal which can rightly be regarded as a portion of conductor.

when both currents were in the same direction, and that they were repelled when the currents were in opposite directions.

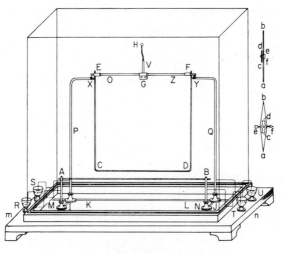

FIG. 44.

Consider now the interaction of an electric current and a magnet and that of two magnets; it will be seen that both come under the same law governing the interaction of two currents, if it is assumed that the current is established at each point of a line drawn on the surface of the magnet from one pole to the other in planes perpendicular to the axis of this magnet; it hardly seems possible to me, from consideration of all the facts, to doubt that such currents do exist about the axis of a magnet, or rather that magnetization is nothing other than the operation by which particles of steel are endowed with the property to produce, in the sense of the current about which we have just been speaking, the same electromotive action as in the voltaic battery, in the oxidized zinc of mineralogists, in heated tourmaline, and even in the battery formed by wet boards and disks of metal at different temperatures. But since with magnets this electromotive action develops between different particles of one and the same body, a good conductor, it can never, as pointed out above, produce electric tension, only a continuous current like that which would occur in a battery connected to itself in a closed curve; it is quite clear from the foregoing observations that such a battery could produce no tensions, attractions, no ordinary electric repulsions, nor any chemical phenomena, since a liquid cannot be interposed in the circuit; it is evident that any current which is established in this battery would immediately act to direct, attract or repel another electric current or a magnet, which, as we shall see, is only an assembly of electric currents.

It is thus that the unexpected result is reached that magnetic phenomena are due solely to electricity and there is no difference between the two poles of a magnet other than their position in regard to the currents of which the magnet is composed, the austral pole† being that to the right of the currents and the boreal pole to their left.

Ever since my first researches on the subject, I have sought to find the law governing the attractive or repulsive action of two electric currents on variation of the distance between them and the angles which determine their position. I was soon convinced that this law could not be found by experiment because no simple representation could be obtained except by considering portions of currents of infinitesimal length, and experiments cannot be performed on such currents; the action of currents with measurable effects is the sum of the infinitesimal actions of the elements, a sum which can only be obtained by two successive integrations, of which one must be performed over the full extent of one current for the same point of the other, whilst the other must be performed with respect to the result of the first between the limits set by the first current over the full extent of the second current; it is only the result of this last integration, taken between the limits set by the extremities of the second current, that can be compared with experimental data; hence, as I said in my dissertation to the Académie on 9 October last, these integrations must be considered before one can determine the interaction of two currents of finite length, whether rectilinear, or curvilinear, bearing in mind that in a curvilinear current the direction of the constituent portions is determined at each point by

---

†The pole which points north in a magnetized needle; it is on the right of the currents forming the magnet because it is to the left of an external current in the same direction facing the needle.

the tangent to the curve which is its path, and that the action of an electric current on a magnet, or between two magnets, is then found by regarding, in these two latter cases, the magnets as assemblies of electric currents arranged in the way I have indicated above. From M. Biot's splendid experiment, currents which are in one and the same plane perpendicular to the axis of a magnet, must be regarded as having the same intensity, since it results from the experiment where he compared the effects produced by the action of the earth on two similarly magnetized bars of the same size and shape, of which one was hollow and the other solid, that the motive force is proportional to the mass and that in consequence the causes to which it is due act with the same intensity on all particles of one and the same cross-section perpendicular to the axis, the intensity varying from section to section according as these sections are close to or far from the poles. When the magnet is a solid of rotation about the line joining its two poles, all the currents of one and the same section must be circles; the calculations for magnets of this shape can be simplified by first calculating the action of an infinitesimal portion of current on an assembly of concentric circular currents occupying the entire space enclosed within the surface of a circle, such that the intensities which are attributed to them in the calculation are proportional to the infinitesimal distance of two consecutive currents measured on their radius (the result of integration would otherwise depend on the number of infinitesimal parts into which this radius were divided by the circumferences representing the currents; which is absurd). Since a circular current is attracted wherever it is in the same direction as a current acting on it, and repelled in the part wherever it is in the opposite direction, the action on the surface of a circle perpendicular to the axis of a magnet consists of a resultant equal to the difference between the components of the attractions and repulsions parallel to this resultant, and of the resultant couple which the attractions and repulsions equally tend to produce. The value of the action is found by integrations with respect to the radii of the surface for a solid magnet, and between the radii of the inside and outside surfaces for a hollow cylinder, and the result of this operation must then be multiplied: (1) by the infinitesimal thickness of the cross-section and the overall intensity of the currents composing it, and (2) by the intensity and the length of the infinitesimal portion of current which is assumed to be acting upon it; the values are thus obtained of the resultant and resultant couple constituting the elemental action between a circular or crown-shaped section and an infinitesimal portion of the current.

Having found this value, if it is a question of the interaction of a magnet and a current, whether curvilinear or rectilinear of finite length, in order to obtain the mutual action, it is only necessary to perform the integrations which are required for calculation of the resultant and resultant couple of all the elemental actions between each section of the magnet and each infinitesimal portion of the current.

But if it is a question of the mutual action of two hollow or solid cylindrical magnets, it is first necessary to obtain the value of the interaction between a circular or crown-shaped section and an infinitesimal portion of current in order to deduce by two integrations the interaction between this section and a similar section (regarding this latter section as composed of circular currents like the first section), the resultant and resultant couple of the mutual action of two infinitely minute sections are thus obtained and by new integrations the same can be obtained with regard to the action of two magnets under the surfaces of rotation, having on each occasion first determined by comparison of the calculated and experimental results the relationship between the distance from each section to one of the magnet poles and the intensity of the section currents. I have still not finished the calculations connected with the action of a magnet on an electric current, nor with the interaction of two magnets†, but only that by which I determined the mutual action

---

†The calculations assume that the presence of an electric current, or of another magnet, changes nothing in the electric currents of a magnet on which they act. This is never the case with soft iron; but since tempered steel preserves the modifications which it undergoes, it seems to me from the experiment of M. Arago with magnetization by electric current and from the procedures of ordinary magnetization, that when magnetized steel is in precisely the same state as prior to the action of another magnet or electric current upon it, it can be inferred that the constituent currents are practically constant in direction or intensity during their action, for otherwise the modifications which they undergo would not persist after the action had ceased.

of two rectilinear currents of finite magnitude, using the hypothesis which agrees best with the observed phenomena and the general results of experiments in respect of the value of the attraction or repulsion which occurs between two infinitesimal portions of electric currents. At first I did not plan to publish this formula or its diverse applications until I had been able to compare it with the results of precise measurement; but, having considered all the circumstances associated with the phenomena, I believed I saw sufficient probability in favour of this hypothesis to give an outline of it now, and this will be the object of the following paragraphs.

I constructed the apparatus shown in Fig. 45 as being more appropriate than my original device for the particular measurements

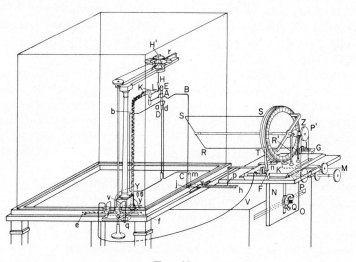

FIG. 45

that I had in mind, especially as the support of the graduated circle, besides its movement which allows the moving conductor to be brought nearer or taken further away, can now also be moved by means of an adjustable screw in two other ways, namely, vertically, and horizontally transverse to the other two movements. The first of these three movements is indispensable for measurement by the device, and originally this was the only possible movement, the aim of the two additional movements being to simplify the measurements when the lines joining the mid-points of the two currents are not perpendicular to them. For this reason I think that adjustment by hand before the experiment is preferable to the use of adjusting screws, provided that the support of the graduated circle can afterwards be fixed in a stable manner in the same position as previously.

The first of the three movements of the support KFG is by the adjusting screw M; the other two movements are by the connecting piece by which the support is fixed to the block of wood N which is free to slide horizontally and vertically on the other block of wood O at the base of the device. A horizontal slot is made in one block and a vertical slot in the other; at the intersection of these two slots there is a screw nut Q which serves to arrest the moving piece on the fixed piece in the desired position. The graduated circle for inclining the attached portion of conducting wire at any designed angle is revolved by the two return pullies P and P'. In order that there should be no action of the earth on the moving conductor to combine with the action of the fixed conductor, the former is made of two equal and opposite parts $ABCd$, $abcDE$ with the shape shown in the diagram; so as to be able to bring its two extremities into contact with the extremities of the battery, the moving conductor is interrupted at the angle A of the suspension piece HH' which balances with torsion the attraction or repulsion of the two currents. The branch BA continues beyond A and DE continues beyond E, both terminating at K and L where the tips are immersed in two small mercury-filled cups without touching the bottom.

There is no need to remind physicists who are accustomed to this type of measurement that owing to continuous variation of current intensity with the energy of the battery, it is necessary to repeat an experiment at some constant distance in between each experiment so as to know how the intensity of the currents varies and its value at each instant from the action observed each time at this constant distance and by the ordinary rules of interpolation. The same approach is to be adopted to compare the attractions and repulsions when the angle between the two currents varies if the line joining their mid-points is constantly perpendicular to them. The intermediate observations are simplified at each instant since, with the distance between the two portions of conductor BC and SR constant, it is sufficient to turn the graduated circle in order to return *SR* each time in the direction parallel to *BC*. Finally, if it is desired to measure the interaction of *BC* and *SR* when the line joining their mid-points is not perpendicular to their direction, the support of the graduated circle is set in the appropriate position by the screw nut *Q* which sets it in the desired position in relation to the rest of the apparatus and then by performing a series of experiments similar to those in the preceding case, the results obtained in each position of the conductors can be compared with those in the case when the line joining the mid-points is perpendicular, this comparison being made for one and the same shorter distance between currents and then for the various other distances; everything necessary is thus obtained to see how and up to what point these different circumstances influence the interaction of the electric currents; it only remains to see if all the results agree with the calculation of the effects which must be produced in each arrangement from the law acknowledged to govern the attraction between two infinitesimal portions of current.

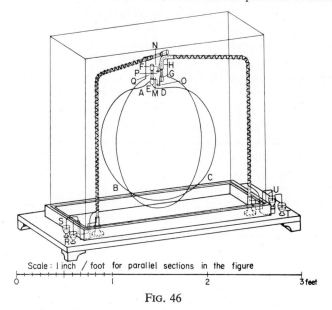

Scale : 1 inch / foot for parallel sections in the figure

FIG. 46

Ampère's diagram of an apparatus for investigating the mutual action of two circular currents.

II. THE INTERACTION BETWEEN AN ELECTRICAL CONDUCTOR AND A MAGNET

This action that M. Oersted discovered led me to look for the interaction of two electrical currents, the action of the earth on a current and the role of electricity in magnetic phenomena in that the distribution in the magnet is similar to that of a conductor with closed curves perpendicular to its axis. These findings, most of which have only recently been confirmed by experiment, were communicated to the Académie in its session of 18 September 1820.

When first I wanted to find the causes of the new phenomena discovered by M. Oersted, I reflected that since the order in which two facts are discovered in no way affects any conclusions which can be drawn from analogies they present, it might, before we knew that a magnetized needle points constantly from south to north, have first been known that a magnetized needle has the property of being influenced by an electric current into a position perpendicular to the current, in such a way that the austral pole of the magnet is carried to the left of the current, and it could then have subsequently been discovered that the extremity of the needle which is carried to the left of the current points constantly towards the north: would not the simplest idea, and the one which would immediately occur to anyone who wanted to explain the constant direction from south to north, be to postulate an electric current in the earth in a direction such that the north would be to the left of a man who, lying on its surface facing the needle, received this current in the direction from his feet to his head, and to draw the conclusion that it takes place from east to west in a direction perpendicular to the magnetic meridian?

Now, if electric currents are the cause of the directive action of the earth, then electric currents could also cause the action of one magnet on another magnet; it therefore follows that a magnet could be regarded as an assembly of electric currents in planes perpendicular to its axis, their direction being such that the austral pole of the magnet, pointing north, is to the right of these currents since it is always to the left of a current placed outside the magnet, and which faces it in a parallel direction, or rather that these currents establish themselves first in the magnet along the shortest closed curves, whether from left to right, or from right to left, and the line perpendicular to the planes of these currents then becomes the axis of the magnet and its extremities make the two poles. Thus, at each pole the electric currents of which the magnet is composed are directed along closed concentric curves; I simulated this arrangement as much as possible by bending a conducting wire in a spiral: this spiral was made from brass wire terminating in two straight portions enclosed in glass tubes† so as to eliminate contact and attach them to the two extremities of the battery.

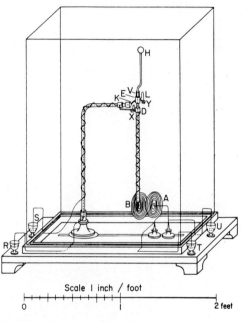

Fig. 47.

Depending on the direction of the current, such a spiral is greatly attracted or repelled by the pole of a magnet which is presented with its axis perpendicular to the plane of the spiral, according as the current of the spiral and of the magnet flow in the same or opposite directions. In replacing the magnet by another spiral with its current in the same direction, the same attractions and repulsions occur;

---

†I have since changed this arrangement as I shall show later.

it is in this way that I discovered that two electric currents attract each other when they flow in the same direction and repel each other in the other case.

Replacing the spirally wound metal wire by another magnet in the experiment on the interaction between the pole of a magnet and the current in a spiral, the effects are still the same, whether in attraction or repulsion, in conformity with the law of the known phenomena of a magnet; it is also evident that all circumstances associated with these phenomena are a necessary corollary of the arrangement of the component electric currents so that they attract or repel each other.

4
The Mémoire Read to the Académie in 1825*

D. AMPÈRE'S PHILOSOPHY OF SCIENCE

On the mathematical theory of electrodynamic phenomena, experimentally deduced, collecting the papers delivered at the Académie Royale des Sciences by M. Ampère on the 4 and 26 December 1820, 10 June 1822, 22 December 1823 and 12 September and 21 November 1825.

The new era in the history of science marked by the works of Newton, is not only the age of man's most important discovery in the causes of natural phenomena, it is also the age in which the human spirit has opened a new highway into the sciences which have natural phenomena as their object of study.

Until Newton, the causes of natural phenomena had been sought almost exclusively in the impulsion of an unknown fluid which entrained particles of materials in the same direction as its own particles; wherever rotational motion occurred, a vortex in the same direction was imagined.

Newton taught us that motion of this kind, like all motions in nature, must be reducible by calculation to forces acting between two material particles along the straight line between them such that the action of one upon the other is equal and opposite to that which the latter has upon the former and, consequently, assuming the two particles to be permanently associated, that no motion whatsoever can result from their interaction. It is this law, now confirmed by every observation and every calculation, which he represented in the three axioms at the beginning of the *Philosophiae naturalis*

Memoires de l'Academie Royale des Sciences 1823 (Issued 1827) (*Translated by* O. M. BLUNN).

principia mathematica. But it was not enough to rise to the conception, the law had to be found which governs the variation of these forces with the positions of the particles between which they act, or, what amounts to the same thing, the value of these forces had to be expressed by a formula.

Newton was far from thinking that this law could be discovered from abstract considerations, however plausible they might be. He established that such laws must be deduced from observed facts, or preferably, from empirical laws, like those of Kepler, which are only the generalized results of very many facts.

To observe first the facts, varying the conditions as much as possible, to accompany this with precise measurement, in order to deduce general laws based solely on experience, and to deduce therefrom, independently of all hypothesis regarding the nature of the forces which produce the phenomena, the mathematical value of these forces, that is to say, to derive the formula which represents them, such was the road which Newton followed. This was the approach generally adopted by the learned men of France to whom physics owes the immense progress which has been made in recent times, and similarly it has guided me in all my research into electrodynamic phenomena. I have relied solely on experimentation to establish the laws of the phenomena and from them I have derived the formula which alone can represent the forces which are produced; I have not investigated the possible cause of these forces, convinced that all research of this nature must proceed from pure experimental knowledge of the laws and from the value, determined solely by deduction from these laws, of the individual forces in the direction which is, of necessity, that of a straight line drawn through the material points between which the forces act. That is why I shall refrain from discussing any ideas which I might have on the nature of the cause of the forces produced by voltaic conductors, though this is contained in the notes which accompany the "Exposé sommaire des nouvelles expériences electromagnétiques faites par plusieurs physiciens depuis le mois de mars 1821," which I read at the public session of the Académie des Sciences, 8 April 1822; my remarks can be seen in these notes on page 215 of my collection of "Observations in Electrodynamics". It does not appear that this approach, the only one which can lead to results which are free of all hypothesis, is preferred by physicists in the rest of Europe like it is by Frenchmen; the famous scientist who first saw the poles of a magnet transported by the action of a conductor in directions perpendicular to those of the wire, concluded that electrical matter revolved about it and pushed the poles along with it, just as Descartes made "the matter of his vortices" revolve in the direction of planetary revolution. Guided by Newtonian philosophy, I have reduced the phenomenon observed by M. Oerstedt, as has been done for all similar natural phenomena, to forces acting along a straight line joining the two particles between which the actions are exerted; and if I have established that the same arrangement, or the same movement of electricity, which exists in the conductor is present also round the particles of the magnets, it is certainly not to explain their action by impulsion as with a vortex, but to calculate, according to my formula, the resultant forces acting between the particles of a magnet and those of a conductor, or of another magnet, along the lines joining the particles in pairs which are considered to be interacting, and to show that the results of the calculation are completely verified by (1) the experiments of M. Pouillet and my own into the precise determination of the conditions which must exist for a moving conductor to remain in equilibrium when acted upon, whether by another conductor, or by a magnet, and (2) by the agreement between these results and the laws which Coulomb and M. Biot have deduced by their experiments, the former relating to the interaction of two magnets, and the latter to the interaction between a magnet and a conductor.

The principal advantage of formulae which are derived in this way from general facts gained from sufficient observations for their certitude to be incontestable, is that they remain independent, not only of the hypotheses which may have aided in the quest for these formulae, but also independent of those which may later be adopted instead. The expression for universal attraction from the laws of Kepler is completely independent of the hypotheses which some writers have advanced to justify the mechanical cause to which they would ascribe it. The theory of heat is founded on general facts which have been obtained by direct observation; the equation deduced from these facts, being confirmed by the agreement between the results of calculation and of experiment, must be equally accepted as representative of the true laws of heat propagation by those who attribute it to the radiation of calorific molecules as by those who take the view that the phenomenon is caused by the vibration of a diffuse fluid in space; it is only necessary for the former to show how the equations results from their way of looking at heat and for the others to derive it from general formulae for vibratory motion; doing so does not add anything to the certitude of the equation, but only substantiates the respective hypotheses. The physicist who refrains from committing himself in this respect, acknowledges the heat equation to be an exact representation of the facts without concerning himself with the manner in which it can result from one or other of the explanations of which we are speaking; and if new phenomena and new calculations should demonstrate that the effects of heat can in fact only be explained in a system of vibrations, the great physicist who first produced the equation and who created the methods of integration to apply it in his research, is still just as much the author of the mathematical theory of heat, as Newton is still the author of the theory of planetary motion, even though the theory was not as completely demonstrated by his works as his successors have been able to do in theirs.

It is the same with the formula by which I represented electrodynamic action. Whatever the physical cause to which the phenomena produced by this action might be ascribed, the formula which I have obtained will always remain the true statement of the facts. If it should later be derived from one of the considerations by which so many other phenomena have been explained, such as attraction in inverse ratio to the square of the distance, considerations which disregard any appreciable distance between particles between which forces are exerted, the vibration of a fluid in space, etc., another step forward will have been made in this field of physics; but this inquiry, in which I myself am no longer occupied, though I fully recognize its importance, will change nothing in the results of my work, since to be in agreement with the facts, the hypothesis which is eventually adopted must always be in accord with the formula which fully represents them.

From the time when I noticed that two voltaic conductors interact, now attracting each other, now repelling each other, ever since I distinguished and described the actions which they exert in the various positions where they can be in relation to each other, and after I had established that the action exerted by a straight conductor is equal to that exerted by a sinuous conductor whenever the latter only deviates slightly from the direction of the former and both terminate at the same points, I have been seeking to express the value of the attractive or repellent force between two elements, or infinitesimal parts, of conducting wires by a formula so as to be able to derive by the known methods of integration the action which takes place between two portions of conductors of the shape in question in any given conditions.

The impossibility of conducting direct experiments on infinitesimal portions of a voltaic circuit makes it necessary to proceed from observations of conductors of finite dimension and to satisfy two conditions, namely that the observations be capable of great precision and that they be appropriate to the determination of the interaction between two infinitesimal portions of wires. It is possible to proceed in either of two ways: one is first to measure values of the mutual action of two portions of finite dimension with the greatest possible exactitude, by placing them successively, one in relation to the other, at different distances and in different positions, for it is evident that the interaction does not depend solely on distance, and then to advance a hypothesis as to the value of the mutual action of two infinitesimal portions, to derive the value of the action which must result for the test conductors of finite dimension, and to modify the hypothesis until the calculated results are in accord with those of observation. It is this procedure which I first proposed to follow, as explained in detail in the paper which I read at the Académie des Sciences 9 October 1820[†]; though it leads to the truth only by the indirect route of hypothesis, it is no less valuable because of that, since it is often the only way open in investigations of this kind. A member of this Académie whose works have covered the whole range of physics has aptly expressed this in the "Notice on the Magnetization of Metals by Electricity in Motion", which he read 2 April 1821, saying that prediction of this kind was the aim of practically all physical research.[‡]

[†]This paper has not been published separately, but the principal results are included in vol. XV of the *Annales de Chimie et de Physique* (1820), (See Part 2, Section 3).

[‡]See *Journal des Savants*, p. 233, April 1821.

However, the same end can be reached more directly in the way which I have since followed: it consists in establishing by experiment that a moving conductor remains exactly in equilibrium between equal forces, or between equal rotational moments, these forces and these moments being produced by portions of fixed conductors of arbitrary shape and dimension without equilibrium being disturbed in the conditions of the experiment, and in determining directly therefrom by calculation what the value of the mutual action of the two infinitesimal portions must be for equilibrium to be, in fact, independent of all variations of shape and dimension compatible with the conditions.

This procedure can only be adopted when the nature of the action being studied is such that cases of equilibrium which are independent of the shape of the body are possible; it is therefore of much more restricted application than the first method which I discussed; but since voltaic conductors do permit equilibrium of this kind, it is natural to prefer the simpler and more direct method which is capable of great exactitude if ordinary precautions are taken for the experiments. There is, however, in connection with the action of conductors, a much more important reason for employing it in the determination of the forces which produce their action: it is the extreme difficulty associated with experiments where it is proposed, for example, to measure the forces by the number of oscillations of the body which is subjected to the actions. This difficulty is due to the fact that when a fixed conductor is made to act upon the moving portion of a circuit, the pieces of apparatus which are necessary for connection to the battery act on the moving portion at the same time as the fixed conductor, thus altering the results of the experiments. I believe, however, that I have succeeded in overcoming this difficulty in a suitable apparatus for measuring the mutual action of two conductors, one fixed and the other moving, by the number of oscillations in the latter for various shapes of the fixed conductor. I shall describe this apparatus in the course of this paper.

It is true that the same obstacles do not arise when the action of a conducting wire on a magnet is measured in the same way; but this method cannot be employed when it is a question of determining the forces which two conductors exert upon each other, the question which must be our first consideration in the investigation of the new phenomena. It is evident that if the action of a conductor on a magnet is due to some other cause than that which produces the effect between two conductors, experiments performed with respect to a conductor and magnet can add nothing to the study of two conductors; if magnets only owe their properties to electric currents, which encircle each of their particles, it is necessary, in order to draw definite conclusions as to the action of the conducting wire on these currents, to be sure that these currents are of the same intensity near to the surface of the magnet as within it, or else to know the law governing the variation of intensity; whether the planes of the currents are everywhere perpendicular to the axis of a bar magnet, as I at first supposed, or whether the mutual action of the currents of the magnet itself inclines them more to the axis when at a greater distance from this axis, which is what I have since concluded from the difference which is noticeable between the position of the poles on a magnet and the position of the points which are endowed with the same properties in a conductor of which one part is helically wound.

E.   AMPÈRE'S BASIC EXPERIMENTS

The various cases of equilibrium which I have established by precise experiment provide the laws leading directly to the mathematical expression for the force which two elements of conducting wires exert upon each other, in that they first make the form of this expression known and then allow the initially unknown constants to be determined, just as the laws of Kepler first show that the force which holds the planets in their orbits tends constantly towards the centre of the sun, since it varies for a particular planet in inverse ratio to the square of its distance to the solar centre, so that the constant coefficient which represents its intensity has the same value for all planets. These cases of equilibrium are four in number: the first demonstrates the equality in absolute value of the attraction and repulsion which is produced when a current flows alternately in opposite directions in a fixed conductor the distance to the body on which it acts remaining constant. This equality results from the simple observation that two equal portions of one and the same conductor which are covered in silk to prevent contact, whether both straight, or twisted together to form round each other two equal helices, in which the same electric current flows, but in opposite directions, exert no action on either a magnet or a moving conductor; this can be established by the moving conductor which is illustrated in Plate I, Fig. 9 of *Annales de Chimie et de Physique* vol. XVIII, relating to the description of the electrodynamic apparatus of mine which is introduced here (Fig. 48).† A horizontal straight conductor

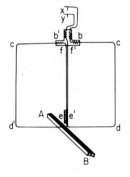

FIG. 48.

AB, doubled several times over, is placed slightly below the lower part *dee'd'* such that its mid-point in length and thickness is in the vertical line through the points *x, y* about which the moving conductor turns freely. It is seen that this conductor stays in the position where it is placed, which proves that there is equilibrium between the actions exerted by the fixed conductor on the two equal and opposite portions of the circuit *bcde* and *b'c'd'e'* which differ only in that the current flows towards the fixed conductor in the one, and away from it in the other, whatever the angle between the fixed conductor and the plane of the moving conductor: now,

---

†The figure is not easy to reproduce clearly. The suspended portion of the apparatus consists of two rectangles so connected that the current circulates round them in opposite directions. The fixed conductor *AB* is, of course, not attached to the moving system.

considering first the two actions exerted between each portion of the circuit and the half of the conductor AB which is the nearest, and then the two actions between each of the two portions and the half of the conductor which is the furthest away, it will be seen without difficulty (1) that the equilibrium under consideration cannot occur at all angles except in so far as there is equilibrium separately between the first two actions and the last two; (2) that if one of the first two actions is attractive because current flows in the same direction along the sides of the acute angle formed by the portions of the conductors,† the other will be repellent because the current flows in opposite directions along the two sides of the equal and opposite angle at the vertex, so that, for there to be equilibrium, the first two actions which tend to make the moving conductor turn, the one in one direction, and the other in the opposite direction, must be equal to each other; and the last two actions, the one attractive and the other repellent, between the sides of the two obtuse and opposite angles at the vertex and the complements of those about which we have just been speaking, must also be equal to each other. Needless to say, these actions are really sums of products of forces which act on each infinitesimal portion of the moving conductor multiplied by their distance to the vertical about which this conductor is free to turn; however, the corresponding infinitesimal portions of the two arms $bcde$ and $b'c'd'e'$ always being at equal distances from the vertical about which they turn, the equality of the moments makes it necessary for the forces to be equal.

The second of the three general cases of equilibrium was indicated by me towards the end of the year 1820; it consists in the equality of the actions exerted on a moving straight conductor by two fixed conductors situated the same distance away from it, of which one is also straight, but the other bent in any manner. This was the apparatus by which I verified the equality of the two actions in the precise experiments, the results of which were communicated to the Académie in the session of 26 December 1820.

---

†*Note by R.A.R.T.* Here Ampère is considering mainly the interaction between the fixed conductor AB and the two pieces of the suspended conductor, de and $d'e'$ which lie nearest to it.

The two wooden posts *PQ*, *RS* (Fig. 49) are slotted on the sides which mutually face each other, the straight wire *bc* being laid in the slot of *PQ*, and the wire *kl* in that of *RS*; over its entire length

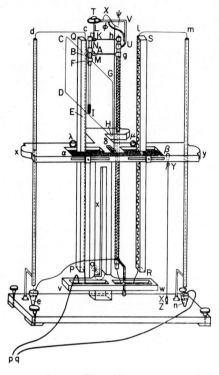

Fig. 49.

this wire is twisted in the plane perpendicular to that joining the two axes of the posts, such that the wire at no point departs more than a very short distance from the mid-point of the slot.

These two wires serve as conductors for the two portions of a current which is made to repel the part *GH* of a moving conductor consisting of two almost closed and equal rectangular circuits *BCDE*, *FGHI* in which the current flows in opposite directions so that the effect of the earth on these two circuits cancels out. At the two extremities of this moving conductor there are two points *A* and *K* which are immersed in the mercury-filled cups *M* and *N* and soldered

to the extremities of the copper arms gM, hN. These arms make
contact via the copper bushings g and h, the first with the copper
wire gfe, helically wound around the glass tube hgf, the other with
the straight wire hi which goes through the inside of this tube to the
trough ki made in the piece of wood vw which is fixed at the desired
height against the pillar z with the set screw o. In view of the experiment to which I referred above, the portion of the circuit composed
of the helix gf and the straight wire hi can exert no action on the
moving conductor. For current to flow in the fixed conductors bc
and kl, the connecting wires of these conductors are continued by
cde, lmn in two glass tubes† attached to the cross-piece xy, finally
terminating, the first in cup e and the other in cup n. The current
flows through the conductors of the apparatus in the following order:
$pabcdefg\,M\,A\,B\,C\,D\,E\,F\,G\,H\,I\,K\,N\,hiklmnq$; as a result, the
current flows up the two fixed conductors and down that part, GH,
of the moving conductor which is acted upon in its position midway
between the two fixed conductors and lies in the plane which passes
through their axes. The part GH is thus repelled by bc and kl,
whence it follows that if the action of these two conductors is the
same at equal distances, GH must remain midway between them;
this is, in fact, what happens.

It is as well to point out (1) that though the two axes of the fixed
conductors are the same distance from GH, it cannot be stated with
rigour that the distance is the same for all points of the conductor kl
owing to its contours and bends. But since these bends are in a plane
perpendicular to the plane through GH and through the axes of the
fixed conductors, it is evident that the resulting difference in distance
is as small as possible and as much less than half of the width of the
slot RS as this half is less than the interval between the two posts
(this difference, in the case when it is the largest possible, is equal
to that between the radius and the secant of an arc with tangent
equal to half the width of the slot and belonging to a circle of which
the diameter is the interval between the two posts); (2) that if each
infinitesimal portion of the conductor kl is resolved in the same way

---

†These tubes are used to prevent flexure of the enclosed wires by holding
them at equal distances from the two conductors bc, kl, so that their
actions on GH, which reduce that of these two conductors, should reduce
them equally.

as a force could be resolved, into two minute portions which are projections, the one along the vertical axis of the conductor and the other along horizontal lines drawn through all points of the conductor in the plane in which it is bent, the sum of the first projections (taking as negative those which, being in the opposite direction, must act in the opposite direction), will be equal to the length of this axis; hence the total action resulting from all these projections is the same as that of a straight conductor equal to the axis, that is to say, it is equal to that of the conductor *bc* situated on the other side at the same distance from *GH*. The other projections will have zero effect on the moving conductor *GH* since the planes erected vertically at the mid-point of each of them pass approximately through *GH*. The joining of these two series of projections thus produces an action on *GH* equal to that of *bc*; and since experience also proves that the sinuous conductor *kl* produces an action equal to that of *bc*, whatever its bends and contours, it follows that it acts in all cases like the combination of the two series of projections, which cannot occur independently of the manner in which the conductor is bent unless each part of it acts separately as the resultant of its two projections.

For this experiment to have the desired exactitude, it is necessary for the two posts to be exactly vertical and at precisely the same distance from the moving conductor. To fulfil these conditions, a support $\alpha\beta$ is matched to the cross-piece *xy* and the posts are fixed by two clamps η, θ and two adjustable screws λ, μ, so that the posts may be moved apart or brought closer together at will, keeping the same distance from the mid-point γ of the support $\alpha\beta$. The apparatus is so constructed that the two posts are perpendicular to the cross-piece *xy*, and this is made horizontal by the screws at the four corners of the base of the device and the plumb line *XY* which corresponds exactly to a point *Z* as conveniently marked on the base with the cross-piece *xy* perfectly level.

For the conductor *ABCDEFGHIK* to revolve about a vertical axis at an equal distance from the two conductors *bc* and *kl*, this conductor is suspended by a very fine metal wire attached to the centre of a knob *T* which can rotate without altering the distance between the two conductors; this knob is at the centre of a small dial *O*, on which the letter *L* marks the place where it is necessary to stop in order that the part *GH* of the moving conductor should hang, without the suspension being twisted, at the mid-point of the interval between the two fixed conductors *bc, kl* in order to be able immediately to return the needle to the position in which it should be whenever it is desired to repeat the experiment. It is checked that *GH* is an equal distance from *bc* and *kl* by another plumb line $\psi\omega$ which is attached to the copper arm $\varphi\alpha\psi$ carried, like the dial *O*, on the support *UVO*, in which this arm is free to revolve about the axis of the knob φ at the end of it, thus making it possible to have the plumb-line ω correspond to the line $\gamma\delta$ in the middle of the support $\alpha\beta$. When the conductor is in the appropriate position, the three verticals $\psi\omega$, *GH* and *CD* are in the same plane, as can easily be checked by placing one's eye in this plane in front of $\psi\omega$.

The moving conductor is thus arranged beforehand in the position where it will be in equilibrium between the repulsions of the two fixed conductors, if these repulsions are exactly equal: these actions are then produced by immersing into the trough *ba* and the cup *n* respectively the wires *ap* and *nq* which connect to the two extremities of the battery, and the conductor *GH* is found to remain in this position despite the great mobility associated with suspensions of this kind. If the mark *L* is displaced, even slightly, which brings *GH* into a position which is no longer equidistant between the fixed conductors *bc, kl*, it is seen to move as soon as communication with the battery is established, swinging away from whichever conductor is the nearest. At the time when I had this device constructed, I established in this way that the actions of the two conductors are equal from sufficient experiments with all the necessary precautions, for there to be no doubt about the result.

The same law can also be demonstrated by a simple experiment for which it is sufficient to take a silk-covered copper wire and to wind a part round the straight portion without being separated from it other than by the silk. It is then found that another portion of the wire does not affect the assembly of two portions; and since it would be the same for an assembly of two straight wires with a similar electric current in opposite directions (from the experiment by which the first case of equilibrium was very simply established), it follows that the action of the current in the wound portion is exactly equal to that of the current in the straight part between identical extremities, because the action of both these two conductors would be counter-balanced by the action of the current in a straight portion of equal length, but in the opposite direction.

The third case of equilibrium is that a closed circuit of any arbitrary shape cannot produce movement in a portion of conducting wire which is in the form of an arc of a circle whose centre lies on a fixed axis about which it may turn freely and which is perpendicular to the plane of the circle of which the arc forms part.

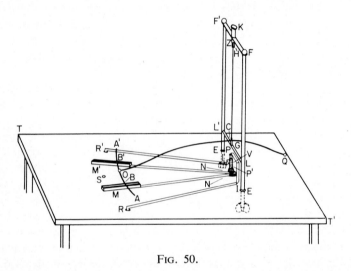

Fig. 50.

On the base table TT' (Fig. 50) two columns EF and $E'F'$ are erected which are joined by the cross-pieces LL', FF'; an upright GH is held in the vertical position between these two cross-pieces. Its two pointed extremities G, H fit into two tapered holes, one in the lower cross-piece LL', the other in the extremity of the screw KZ carried by the upper traverse FF' which locates the upright GH without locking it. At C a support QO is fixed rigidly to this upright. At its extremity O is a hinge which engages the mid-point of the circular arc AA' (formed by a metal wire) which remains constantly in the horizontal position and the distance from the point O to the axle GH in radius. This arc is held in equilibrium by the counter-weight Q, thus reducing the friction of the upright GH in the tapered holes where its extremities are held. Below the arc AA' there are two small troughs M, M' which are filled with mercury so that the surface of the mercury, rising above the brim, just touches the arc AA' at B and B'. These two small troughs are connected by the metallic conductors MN, $M'N'$ to the cups P, P', which are full of mercury. The cup P and the conductor MN, which connects it to the trough M, are fixed to a vertical upright which is bedded in the table, but leaving it free to turn. The cup P', to which the conductor $M'N'$ is connected, is traversed by the same upright, about which it, too, can revolve independently. The cup is insulated from the upright by the glass tube V which envelopes it, and by the glass ring U which separates it from the conductor of the trough M so as to be able to arrange the conductors MN, $M'N'$ at any desired angle.

Two other conductors IR and $I'R'$, attached to the table, are immersed respectively in the cups P and P' and connect them to the cavities R, R' which are made in the table and filled with mercury. Finally, a third cavity S, likewise full of mercury, is situated in between the other two.

This apparatus is used in the following way: one of the battery wires, say, the positive wire, is immersed in the cavity R, whilst the negative is immersed in S, which is made to communicate with the cavity R' by a curvilinear conductor of arbitrary shape. The current follows the conductor RI, passes into the cup P, and thence to the conductor MN, the trough M, the conductor $M'N'$, the cup P', the conductor $I'R'$ and finally from the cavity R' into the curvilinear conductor which connects to the mercury of the cavity S, where the negative wire of the battery is immersed.

With this arrangement the total circuit is formed by:

(1) the arc BB' and the conductors MN, $M'N'$;
(2) the circuit consisting of the parts RIP and $P'I'R'$ of the apparatus, the curvilinear conductor from R' to S and the battery itself.

This latter circuit must act as a closed circuit since it is only interrupted by the glass which insulates the two cups P, P'; it is therefore sufficient to observe its action on the arc BB' to determine the action of a closed circuit on an arc in different positions in relation to each other.

When by means of the hinge O the arc AA' is positioned such that its centre lies outside the axis GH, the arc moves and slides on the mercury of the troughs M, M' owing to the action of the closed curvilinear current flowing from R' to S. If, however, its centre is on the upright, it remains stationary; hence, the two portions of the closed circuit which tend to make it turn in opposite directions about the axis, exert torques on this arc which are equal in absolute value, no matter what the magnitude of the part BB', as determined by the opening of the angle of the conductors MN, $M'N'$. If, therefore, two arcs BB' are taken in succession which hardly differ from each other, since the torque is zero for both of them, it will also be zero for the slight difference between them, and, in consequence, it is likewise zero for any element of circumference with centre on the axis; hence the direction of the action exerted by the closed circuit on the element is along the upright and it is necessarily perpendicular to the element.

When the arc AA' is positioned so that its centre is on the upright, the conductors MN, $M'N'$ exert equal, but opposite, repulsion on the arc BB' with the result that no effect is produced; since no movement occurs, it is certain that no moment of rotation is produced by the closed circuit.

When the arc AA' moves in the other situation which we envisaged, the actions of the conductors MN and $M'N'$ are no longer equal; it could be thought that the movement was due solely to this difference if the movement did not increase, or decrease, according as the curvilinear circuit from R' to S comes nearer or moves further away, which leaves no doubt that the closed circuit plays a prominent part in the effect.

This result, occurring for any length of the axis AA', will necessarily occur for each of the elements of which the arc is composed. The general conclusion may therefore be drawn that the action of a closed circuit, or of an assembly of closed circuits, on an infinitesimal element of an electric current is perpendicular to this element.

It is by the fourth case of equilibrium, about which I have still to speak, that the constant coefficients occurring in my formula may be finally determined without recourse, as I first had to have, to experiments where a magnet and a conductor interact. The device by which this determination may rest solely on observation of the interaction of two conductors is shown in Fig. 51.

A cavity A is made in the table MN. The cavity is filled with mercury and from it runs the fixed conductor $ABCDEFG$ made from a sheet of copper. The part CDE is circular, and the parts CBA and EFG are insulated from each other by a silk covering.

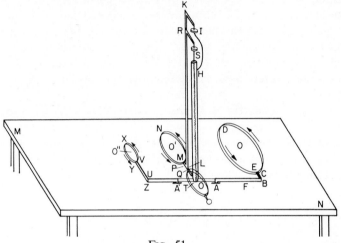

Fig. 51.

At G this conductor is soldered to the copper tube GH, which carries the cup I which is in contact with the tube by means of the copper support HI. The moving conductor $IKLMNPQRS$, of which the part MNP is circular, starts from the cup I; the parts MLK and PQR are insulated by a silk covering. The conductor is held horizontal by the counterweight a fixed on the circumference of a circle formed round the tube GH by the continuation bcg of the sheet constituting the moving conductor. The cup S is supported by the rod ST which has the same axis as GH, but from which it is insulated by a resinous substance which is poured into the tube. The base of the rod ST is soldered to the fixed conductor $TUVXYZA'$, which passes out of the tube GH through an opening large enough for the resin to insulate it as completely at this place as in the rest of the tube GH with regard to ST. At the outlet from the tube this conductor is covered with silk to prevent contact between the portions TUV and

YZA'. The portion VXY is circular and the extremity A' is immersed in the mercury-filled cavity A' in the table.

The centres O, O', O'' of the three circular portions are in a straight line; the radii of the circles which they form are in continuous geometric proportion and the moving conductor is first placed in such a way that the distances OO', $O'O''$ bear the same relation to each other as consecutive terms in this proportion; hence the cirles O and O' form a system similar to that of the circles O' and O''. The positive battery wire is then immersed in A with the negative in A', and the current flows in succession through the circles with centres at O, O', O'', which repel each other in pairs, because the current flows in the opposite direction in neighbouring parts.

The purpose of the experiment is to prove that the moving conductor remains in equilibrium in the position where the ratio of OO' to $O'O''$ is the same as that of the radii of two consecutive circles, and that if it is moved away from this position, it returns to it after oscillating about it.

F.   AMPÈRE'S THEORY OF CURRENT ELEMENTS

I will now explain how to deduce rigorously from these cases of equilibrium the formula by which I represent the mutual action of two elements of voltaic current, showing that it is the only force which, acting along the straight line joining their mid-points, can agree with the facts of the experiment. First of all, it is evident that the mutual action of two elements of electric current is proportional to their length; for, assuming them to be divided into infinitesimal equal parts along their lengths, all the attractions and repulsions of these parts can be regarded as directed along one and the same straight line, so that they necessarily add up. This action must also be proportional to the intensities of the two currents. To express the intensity of a current as a number, suppose that another arbitrary current is chosen for comparison, that two equal elements are taken from each current, and that the ratio is required of the actions which they exert at the same distance on a similar element of any other current if it is parallel to them, or if its direction is perpendicular to the straight lines which join its mid-point with the mid-points of two other elements. This ratio will be the measure of the intensity of one current, assuming that the other is unity.

Let us put i and i' for the ratios of the intensities of two given currents to the intensity of the reference current taken as unity, and put ds and ds' for the lengths of the elements which are considered in each of them; their mutual action, when they are perpendicular to the line joining their mid-points, parallel to each other and situated a unit distance apart, is expressed by $i\,i'\,ds\,ds'$; we shall take the sign $+$ when the two currents, flowing in the same direction, attract, and the sign $-$ in the other case.

If it is desired to relate the action of the two elements to gravity, the weight of a unit volume of suitable matter could be taken for the unit of force. But then the current taken as unity would no longer be arbitrary; it would have to be such that the attraction between two of its elements ds, ds', situated as we have just said, could support a weight which would bear the same relation to the unit of weight as ds, ds' bears to 1. Once this current were determined, the product $i\,i'\,ds\,ds'$ would denote the ratio of the attraction of two elements of arbitrary intensity, still in the same situation, to the weight which would have been selected as the unit of force.

Suppose we now consider two elements placed arbitrarily; their mutual action will depend on their lengths, on the intensities of the currents of which they are part, and on their relative position. This position can be determined by the length r of a straight line joining their mid-points, the angles θ and θ' between a continuation of this line and the directions of the two elements in the same direction as their respective currents, and finally by the angle ω between the planes drawn through each of these directions and the straight line joining the mid-points of the elements.

Consideration of the diverse attractions and repulsions observed in nature led me to believe that the force which I was seeking to represent, acted in some inverse ratio to distance; for greater generality, I assumed that it was in inverse ratio to the nth power of this distance, n being a constant to be determined. Then, putting ρ for the unknown function of the angles θ, θ', ω, I had $\rho\,i\,i'\,ds\,ds'/r^{n}$ as the general expression for the action of two elements ds, ds' of the two currents with intensity i and i' respectively. It remained to

determine the function ρ. For that I shall first consider two elements
ad, a'd' (Fig. 52), parallel to each other, perpendicular to the straight
line joining their mid-points, and a distance r apart; their action
being represented in accordance with the foregoing remarks by
i i' ds ds/r^n, I assumed that ad remained fixed and that a'd' was
transported parallel to itself in such a way that its mid-point was

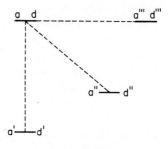

FIG. 52.

always the same distance from the mid-point of ad; ω being always
zero, the value of their mutual action could depend only on the
angles represented above by θ, θ' and which, in this case, are equal, or
complements of each other, according as the currents flow in the same
or opposite direction; in this way I obtained the value i i' ds ds'φ(θ,θ')/r^n.
By putting k for the positive or negative constant to which φ (θ, θ')
is reduced when the element a'd' is at a'''d''' on the continuation
of ad and in the same direction, I obtained k ii' ds ds'/r^n to represent the action of ad on a'''d'''; in this expression the constant
k represents the ratio of the action of ad on a'''d''' to that of
ad on a'd', a ratio which is independent of the distance r, the intensities i, i' and of the lengths ds, ds' of the two elements under
consideration.

These values of the electrodynamic action are sufficient, in the two
simplest cases, for finding the general form of the function ρ by
reason of the experiment, which shows that the attraction of an
infinitely small rectilinear element is the same as that of any other
sinuous element, terminating at the ends of the first, and the theorem
which I have just established, namely that an infinitely small portion
of current exerts no action on another infinitesimal portion of a

current which is situated in a plane which passes through its mid-point and which is perpendicular to its direction. In fact, the two halves of the first element produce equal actions on the second, the one attractive and the other repellent, because the current tends to approach the common perpendicular in one of these halves and to move away from it in the other. These two equal forces form an angle which tends to two right angles according as the element tends to zero. Their resultant is therefore infinitesimal in relation to these forces and in consequence it can be neglected in the calculations. Let

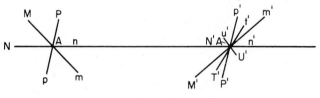

FIG. 53.

$Mm = ds$ and $M'm' = ds'$ (Fig. 53) represent two elements of electric currents with mid-points at A and A'; suppose that the plane $MA'm$ passes along the straight line AA' which joins them, and through the element Mm. We replace the portion of current ds which flows through this element by its projection $Nn = ds \cos \theta$ on the straight line AA' and its projection $Pp = ds \sin \theta$ on the perpendicular erected at A to this straight line in the plane $MA'm$; we then replace the portion of current ds' which flows through $M'm'$ by its projection $N'n' = ds' \cos \theta$ on the straight line AA' and its projection $P'p' = ds' \sin \theta'$ on the perpendicular to AA' drawn through the point A' on AA' in the plane $M'Am'$; finally, we replace the latter by its projection $T't' = ds' \sin \theta' \cos \omega$ in the plane $MA'm$ and its projection $U'u' = ds' \sin \theta' \sin \omega$ on the perpendicular to this plane through the point A'; according to the foregoing law, the two elements ds and ds' exert the same action as the two portions of current $ds \cos \theta$ and $ds \sin \theta$ exert together on the three portions

$$ds' \cos \theta', \ ds' \sin \theta' \cos \omega, \ ds' \sin \theta' \sin \omega;$$

since the latter has its mid-point in the plane MAm to which it is perpendicular, no action occurs between it and the two portions $ds \cos \theta$, $ds \sin \theta$ which are in this plane. For the same reason, there can be no action between the portions $ds \cos \theta$, ds' sin θ', nor between the portions ds sin θ, ds' cos θ, since, imagining a plane through the straight line AA' perpendicular to the plane $MA'm$, ds cos θ and ds' cos θ' are in this plane and the portions ds' sin θ' cos ω and ds sin θ are perpendicular to it with their midpoints in this same plane. The action of the two elements ds and ds' therefore reduces to the two joint remaining actions, namely the interaction between ds sin θ, and ds' sin θ' cos ω and between ds cos θ, and ds' cos θ, these two actions both being along the straight line AA' joining the mid-points of the currents between which they are exerted, and it suffices to add these to obtain the mutual action of the two elements ds and ds'. Now the portions ds sin θ and ds' sin θ' cos ω are in one and the same plane and both are perpendicular to the straight line AA'; accordingly, their mutual action along this straight line is

$$\frac{ii' \, \mathrm{d}s \, \mathrm{d}s' \sin \theta \sin \theta' \cos \omega}{r^n}$$

and that of the two portions ds cos θ and ds' cos θ' along the same line is

$$\frac{ii' \, k \, \mathrm{d}s \, \mathrm{d}s' \cos \theta \cos \theta'}{r^n},$$

thus the interaction of the two elements ds, ds' is necessarily represented by

$$\frac{ii' \, \mathrm{d}s \, \mathrm{d}s'}{r^n} \, (\sin \theta \sin \theta' \cos \omega + k \cos \theta \cos \theta').$$

This formula is simplified by introducing ε for the angle between the two elements in place of ω; for, by considering the spherical triangle with sides θ, θ', ε, we have

$$\cos \varepsilon = \cos \theta \cos \theta' + \sin \theta \sin \theta \cos \omega;$$

hence

$$\sin \theta \sin \theta' \cos \omega = \cos \varepsilon - \cos \theta \cos \theta';$$

substituting this in the foregoing formula and putting $k-1=h$, we get

$$\frac{ii' \mathrm{d}s \, \mathrm{d}s'}{r^n} \, (\cos \varepsilon + h \cos \theta \cos \theta').$$

It is as well to point out that a change of sign occurs when one of the currents, say that of the element ds, takes the diametrically opposite direction, for at that time $\cos \theta$ and $\cos \varepsilon$ change sign, whilst $\cos \theta'$ remains the same. This value of the mutual action of the two elements has only been obtained by the substitution of projections for the element itself; but it may be inferred without difficulty that an element can be replaced by some polygonal contour, or by some curve which terminates at the same extremities, provided that all the dimensions of this polygon or curve are infinitesimal.

Suppose, in fact, that ds_1, ds_2, ..., ds_m are different sides of the infinitesimal polygon which is substituted for ds; AA' may always be regarded as in the same direction as the lines joining the respective mid-points of the sides with A'.

Let θ_1, θ_2, ..., θ_m be the angles which they form respectively with AA', and ε_1, ε_2, ..., ε_m be those which they form with $M'_m{}'$; using Σ to denote a sum of terms of like form, the sum of the actions of the sides ds_1, ds_2, ..., ds_m on ds' is

$$\frac{ii'\,ds'}{r^n} \left(\Sigma\, ds_1 \cos \varepsilon_1' + h \cos \theta' \,\Sigma\, ds_1 \cos \theta_1 \right).$$

Now $\Sigma\, ds_1 \cos \varepsilon_1$ is the projection of the polygonal contour on the direction of ds' and, in consequence, it is equal to the projection of ds on the same direction, that is to say, it is equal to $ds \cos \varepsilon$; likewise $\Sigma\, ds_1 \cos \theta_1$ is equal to the projection of ds on AA' which is $ds \cos \theta$; the action exerted on ds' by the polygonal contour terminated at the extremities of ds may therefore be represented as

$$\frac{ii'\,ds'}{r^n} \left(ds \cos \varepsilon + h\, ds \cos \theta \cos \theta' \right)$$

and it is the same as that of ds on ds'.

Since this conclusion is independent of the number n of sides ds_1, ds_2, ..., ds_m, it also applies to an infinitesimal arc of a curve.

It could likewise be proved that the action of ds' on ds can be replaced by that which an infinitesimal curve, having the same extremities as ds', would exert on each element of the small curve which we have just substituted for ds, and which would therefore be exerted on this small curve itself. Thus, the formula which we have obtained expresses the fact that a curvilinear element produces the same effect as an infinitesimal portion of rectilinear current with the same extremities, whatever the values of the constants n and h. The experiment by which this result has been reached cannot therefore help in the determination of these constants.

We shall therefore have to utilize two of the other cases of equilibrium which we have discussed. But first we shall transform the foregoing expression for the action of two elements of voltaic currents by introducing the partial differentials of the distance of these two elements.

Let x, y, z be the coordinates of the first point, and x', y', z' those of the second. We get:

$$\cos \theta = \frac{x-x'}{r}\frac{dx}{ds} + \frac{y-y}{r}\frac{dy}{ds} + \frac{z-z'}{r}\frac{dz}{ds},$$

$$\cos \theta' = \frac{x-x'}{r}\frac{dx'}{ds} + \frac{y-y'}{r}\frac{dy'}{ds'} + \frac{z-z'}{r}\frac{dz'}{ds'},$$

but since

$$r^2 = (x-x')^2 + (y-y')^2 + (z-z)^2,$$

by successively taking the partial differential coefficients with respect to s and s',

$$r\frac{dr}{ds} = (x-x')\frac{dx}{ds} + (y-y')\frac{dy}{ds} + (z-z')\frac{dz}{ds},$$

$$r\frac{dr}{ds'} = -(x-x')\frac{dx'}{ds'} - (y-y')\frac{dy'}{ds'} - (z-z')\frac{dz'}{ds'},$$

therefore

$$\cos \theta = \frac{dr}{ds}, \quad \cos \theta' = -\frac{dr}{ds'}.$$

To obtain the value of $\cos \varepsilon$, note that

$$\frac{dx}{ds}, \quad \frac{dy}{ds}, \quad \frac{dz}{ds}, \quad \text{and} \quad \frac{dx'}{ds'}, \quad \frac{dy'}{ds'}, \quad \frac{dz'}{ds'}$$

are the cosines of the angles formed by ds and ds' with the three axes, and it follows that

$$\cos \varepsilon = \frac{dx}{ds} \cdot \frac{dx'}{ds'} + \frac{dy}{ds} \cdot \frac{dy'}{ds'} + \frac{dz}{ds} \cdot \frac{dz'}{ds'}.$$

Now, differentiating with respect to s' the foregoing equation which gives $r\, dr/ds$, it is found that

$$r\frac{d^2r}{ds\, ds'} + \frac{dr}{ds} \cdot \frac{dr}{ds'} = -\frac{dx}{ds} \cdot \frac{dx'}{ds'} - \frac{dy}{ds} \cdot \frac{dy'}{ds'} - \frac{dz}{ds} \cdot \frac{dz'}{ds'} = -\cos \varepsilon.$$

If in the formula for the mutual action of two elements ds, ds' we substitute for $\cos \theta$, $\cos \theta$, $\cos \varepsilon$ the values which have just been obtained, and putting $k=1+h$, the formula for the mutual action of the two elements ds, ds' becomes,

$$-\frac{ii'\, ds\, ds'}{r^n} \left(r\frac{d^2r}{ds\, ds'} + k\frac{dr}{ds} \cdot \frac{dr}{ds'} \right),$$

which can be written as

$$-\frac{ii'\, ds\, ds'}{r^n} \cdot \frac{1}{r^{k-1}} \cdot \frac{d\,(r^k\, dr/ds)}{ds'},$$

or

$$ii'\, r^{1-n-k} \frac{d\,(r^k\, dr/ds)}{ds'}\, ds\, ds',$$

It could also be given the following form:

$$-\frac{ii'}{1+k}\, r^{1-n-k}\, \frac{d^2(r^{1+k})}{ds\, ds'}\, ds\, ds'.$$

Let us now examine the result of the third case of equilibrium which shows that the component of the action of a closed circuit on an element in the same direction as this element is always zero, whatever the form of the circuit. Putting ds' for the element in question, the action of an element ds of the closed circuit on ds' is, according to the foregoing,

$$-ii'\,ds'.\,r^{1-n-k}\,\frac{d\,(r^k\,dr/ds')}{ds}\,ds,$$

or, substituting $-\cos\theta'$ for dr/ds',

$$ii'ds'\,r^{1-n-k}\,\frac{d(r^k\cos\theta')}{ds}\,ds;$$

the component of this action along ds' is obtained by multiplying this expression by $\cos\theta'$:

$$ii'\,ds'\,r^{1-n-k}\cos\theta'\,\frac{d\,(r^k\cos\theta')}{ds}\,ds.$$

This differential, integrated over the circuit s, yields the total tangential component and it must be zero whatever the form of the circuit. Integrating it by parts, having written it thus

$$ii'\,ds'\,r^{1-n-2k}r^k\cos\theta'\,\frac{d\,(r^k\cos\theta')}{ds}\,ds,$$

we shall have

$$\tfrac{1}{2}\,ii'\,ds'\left[r^{1-n}\cos{}^2\theta-(1-n-2k)\int r^{-n}\cos^2\theta'dr\right].$$

The first term $r^{1-n}\cos^2\theta'$ vanishes at the limits. As for the integral $\int r^{1-n}\cos^2\theta'dr$, it is very easy to imagine a closed circuit for which it does not reduce to zero. In fact, if this circuit is cut by very close spherical surfaces with centre at the mid-point of the element ds', the two points at which each of these spheres cuts the circuit, give the same value for r and equal values and opposite signs for dr; but the values of $\cos^2\theta'$ may be different and the squares of all the cosines corresponding to the points situated on one side of the extreme points of the circuit may be made less than those relative to the corresponding points on the other side in an infinite number of ways; now, in this case, the integral does not vanish; and as the above expression must be zero, whatever the form of the circuit, the coefficient $1-n-2k$ of this integral must therefore be zero, which gives our first relation between n and k

$$1-n-2k=0.$$

Note by R.A.R.T.

[The derivation of Ampère's formula is virtually complete at this point. The value of n may be determined very simply by the method of dimensions, utilizing the results of the fourth experiment which shows that the forces exerted by one circuit on another is independent of the linear dimensions of the circuits. In the expression

$$\frac{i\,i'\,ds\,ds'}{r^n}(\sin\theta\sin\theta'\cos\omega + k\cos\theta\cos\theta')$$

the numerator is of two dimensions in length. It follows that the denominator must also possess two dimensions in length and thus the value of the exponent n must be 2. This then gives $k=-\frac{1}{2}$.

Apparently Ampère did not discover this line of reasoning until after the writing of his Mémoire and it is given only in a note at the end. The method occurred to him after considering similar conclusions drawn by Laplace from some of Biot's experiments.]

G.   AMPÈRE'S QUANTITATIVE THEORY OF MAGNETISM

Until now we have considered the mutual action of currents in the same plane and rectilinear currents situated arbitrarily in space; it still remains to consider the mutual action of curvilinear currents which are not in the same plane. First we shall assume that these currents describe planar and closed curves with all their dimensions infinitesimal. As we have seen, the action of a current of this kind depends on the three integrals†:

$$A=\lambda\left(\frac{\cos\xi}{l^3}-\frac{3\,q\,x}{l^5}\right),$$

$$B=\lambda\left(\frac{\cos\eta}{l^3}-\frac{3qy}{l^5}\right),$$

$$C=\lambda\left(\frac{\cos\zeta}{l^3}-\frac{3\,q\,z}{l^5}\right).$$

†*Note by R.A.R.T.* Ampère's own derivation of these integrals is rather cumbersome and has been omitted. A more concise derivation is to be found in the commentary, Chapter III, p. 78 et seq.

Imagine a line in space *MmO* (Fig. 54) which the currents encircle, forming very small closed circuits in infinitely close planes which are perpendicular to this line such that the areas λ occupied by these circuits are all equal, that their centres of gravity are on *MmO*, and that these planes are equally spaced along the line. Putting *g* for the

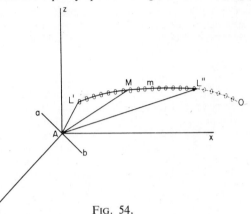

FIG. 54.

infinitesimal distance between neighbouring planes, the number of currents corresponding to an element d*s* of the line *MmO* is d*s*/*g*. It is necessary to multiply the values of *A*, *B*, *C* which have just been found for a single circuit by this number so as to obtain the values which refer to the circuits of the element d*s*. By integrating over the arc *s* from the one extremity *L'* to the other *L''*, values of *A*, *B*, *C* are obtained for the set of circuits which encircle it, an assembly which I have called an *electrodynamic solenoid*, from the Greek word σωληνοειδηζ, which means that which is as a *canal* (pipe), that is to say, it connotes the cylindrical form of the circuits.

Thus, for the solenoid

$$A = \frac{\lambda}{g} \int \left(\frac{\cos \xi \, ds}{l^3} - \frac{3 q x \, ds}{l^5} \right),$$

$$B = \frac{\lambda}{g} \int \left(\frac{\cos \eta \, ds}{l^3} - \frac{3 q y \, ds}{l^5} \right),$$

$$C = \frac{\lambda}{g} \int \left(\frac{\cos \zeta \, ds}{l^3} - \frac{3 q z \, ds}{l^5} \right).$$

Now, since the line g which is perpendicular to the plane of λ, is parallel to the tangent to the curve s, it follows that

$$\cos \xi = \frac{dx}{ds}, \ \cos \eta = \frac{dy}{ds}, \ \cos \zeta = \frac{dz}{ds}.$$

Moreover, q is evidently equal to the sum of the projections of the three coordinates x, y, z on its direction; thus

$$q = \frac{xdx+ydy+zdz}{ds} = \frac{ldl}{ds},$$

since $l^2 = x^2+y^2+z^2$. Substituting these values into the expression which has just been found for C, it becomes†

$$C = \frac{\lambda}{g} \int \left(\frac{dz}{l^3} - \frac{3\,z\,dl}{l^4} \right) = \frac{\lambda}{g} \left(\frac{z}{l^3} + C \right).$$

Putting x', y', z', l' and x'', y'', z'', l'' for the respective values of x, y, z, l at the two extremities L', L'' of the solenoid, we have

$$C = \frac{\lambda}{g} \left(\frac{z''}{l''^3} - \frac{z'}{l'^3} \right).$$

Likewise, finding similar expressions for the two other integrals A, B, the values for the three quantities which it is proposed to calculate for the entire solenoid are:

$$A = \frac{\lambda}{g} \left(\frac{x''}{l''^3} - \frac{x'}{l'^3} \right), \qquad B = \frac{\lambda}{g} \left(\frac{y''}{l''^3} - \frac{y'}{l'^3} \right),$$

$$C = \frac{\lambda}{g} \left(\frac{z''}{l''^3} - \frac{z'}{l'^3} \right).$$

For a solenoid with a closed curve as its directrix, $x''=x'$, $y''=y'$, $z''=z'$, $l''=l'$ and therefore $A=0$, $B=0$, $C=0$; if they extend to infinity in both directions, all the terms of the values of A, B, C will be zero separately, and it is evident that in these two cases the

---

†*Note by R.A.R.T.* Ampère uses the same symbol C, for the arbitrary constant of integration and the expression on the left hand side of the equation. The two are, of course, not the same.

action exerted by the solenoid will be reduced to zero. Assuming that it only extends to infinity on one side, which I shall indicate by referring to it as a semi-infinite solenoid, it is only necessary to consider the extremity with coordinates x' y', z' of finite value, because the other extremity is assumed to be infinitely remote and the first terms of the values which have just been found for A, B, C are necessarily zero; thus

$$A= -\frac{\lambda x'}{gl'^3}, \quad B= -\frac{\lambda y'}{gl'^3}, \quad C= -\frac{\lambda z'}{gl'^3},$$

and therefore $A : B : C : : x' : y' : z'$; hence the normal to the directing plane which passes through the origin and forms angles to the axes with cosines

$$\frac{A}{D}, \quad \frac{B}{D}, \quad \frac{C}{D}$$

where $D=\sqrt{(A^2+B^2+C^2)}$, also passes through the extremity of the solenoid with the coordinates x', y', z'.

As we have seen, in the general case, the total resultant is perpendicular to this normal;† thus the action of an indefinite solenoid on an element is perpendicular to the straight line joining the mid-point of this element to the extremity of the solenoid; and since it is likewise perpendicular to the element, it follows that it is also perpendicular to the plane drawn through this element and through the extremity of the solenoid.

Its direction being determined, it only remains to find its value; now, according to the analysis for the general case, this value is

$$- \frac{D \, ii' \, ds' \sin \varepsilon'}{2},$$

where ε' is the angle between the element ds' and the normal to the

---

†*Note by R.A.R.T.* Ampère's directrix possesses the same direction as what is now called the magnetic induction. This can be seen from the expression for the magnetic field deduced on p. 79.

directing plane; and since $D = \sqrt{(A^2 + B^2 + C^2)}$, it is found without difficulty that

$$D = -\frac{\lambda}{gl'^2},$$

which gives for the value of the resultant

$$\frac{\lambda\, ii'\, \mathrm{d}s' \sin \varepsilon}{2gl'^2}.$$

It is therefore seen that the action exerted by an indefinite solenoid with its extremity at L' (Fig. 54) on the element ab is normal at A to the plane bAL', proportional to the sine of the angle of bAL', and in inverse ratio to the square of the distance AL', and it always remains the same whatever the shape and direction of the indefinite curve $L'L''$ on which all the centres of gravity of the currents composing the indefinite solenoid are assumed to lie.

If it should be desired to consider a definite solenoid with its two extremities situated at two given points L', L'', it is sufficient to assume a second indefinite solenoid commencing at the point L'' of the first and coinciding with it from this point to infinity, with currents opposite in direction, but equal in intensity, the action of the latter being opposite in sign to that of the first indefinite solenoid from L', and destroying its action over the part extending from L'' to infinity in the direction $L''O$ where they are superposed. The action of the solenoid $L'L''$ will therefore be the same as that which would be exerted by joining the two indefinite solenoids and, in consequence, it consists of the force which has just been calculated and another force which acts in the opposite direction, passing likewise through point A, perpendicular to the plane bAL'', and having for value

$$\frac{\lambda\, ii'\, \mathrm{d}s' \sin \varepsilon''}{2gl''^2},$$

where ε'' is the angle bAL'', and l'' is the distance AL''. The total action of the solenoid is the resultant of these two forces and, like them, it passes through point A.

Since the action of a definite solenoid can be deduced directly from that of an indefinite solenoid, we shall in all that remains to be said on the subject proceed from the indefinite solenoid. This simplifies the calculations and conclusions can readily be drawn for a definite solenoid.

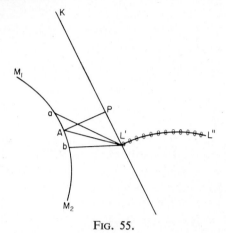

FIG. 55.

Let L' (Fig. 55) be the extremity of an indefinite solenoid, A the mid-point of an element ba of the current $M_1\,AM_2$, and $L'K$ a fixed straight line through the point L'; we put θ for the variable angle $KL'A$, μ for the inclination of the planes bAL', $AL'K$ to each other, and l' for the distance $L'A$. Since the action of the element ba on the solenoid is equal and opposite to that which the solenoid exerts on the element, for its determination it is necessary to consider the mid-point of A which is permanently associated with the solenoid and which is influenced by a force which, ignoring the sign, may be represented as

$$\frac{\lambda\,ii'\,\mathrm{d}s'\sin b\,A\,L'}{2gl'^2} \quad \text{or} \quad \frac{\lambda\,ii'\,\mathrm{d}\nu}{gl'^3},$$

where $\mathrm{d}\nu$ is the area $aL'b$ equal to

$$\frac{l'\,\mathrm{d}s'\sin b\,A\,L'}{2}.$$

This force being normal at A to the plane $AL'b$, to obtain its moment about the axis $L'K$, it is necessary to find the component which is perpendicular to $AL'K$ and to multiply it by a perpendicular to AP

dropped from point A on to the straight line $L'K$. Since μ is the angle between the planes $AL'B$, $AL'K$, this component is obtained by multiplying the foregoing expression by $\cos \mu$; but $d\nu \cos \mu$ is the projection of the area $d\nu$ on the plane $AL'K$, whence it follows that in representing this projection by du, the value of the required component is

$$\frac{\lambda \, ii' \, du}{gl'^3}.$$

Now, the projection of the angle $aL'b$ on $AL'K$ can be regarded as the infinitesimal difference between the angles $KL'a$ and $KL'b$; it is therefore $d\theta$ and we have

$$du = \frac{l'^2 \, d\theta}{2};$$

which reduces the previous expression to

$$\frac{\lambda \, ii' \, d\theta}{2 \, gl'};$$

but since $AP = l' \sin \theta$, for the required moment we have

$$\frac{\lambda ii'}{2 \, g} \sin \theta \, d\theta.$$

This expression, integrated over the curve M_1AM_2, yields the moment of the current making the solenoid revolve about $L'K$: now, if the current is closed, the integral, which in general is

$$C - \frac{\lambda \, ii' \cos \theta}{2g},$$

vanishes between the limits and the moment is zero in respect of any straight line $L'K$ through the point L'.

Hence, in the action of a closed circuit, or of any system of closed circuits, on an indefinite solenoid, all the forces applied to the various elements of the system produce the same moments about the axis as if they were at the extremity of the solenoid; their resultant passes through this extremity and in no case can the forces tend to impart rotational motion to the solenoid about a straight line through its extremity, which is in agreement with the results of the experiments. If the current represented by the curve M_1AM_2 were not closed, its moment for rotation of the solenoid about $L'K$, putting θ_1' and θ_2' for the extreme values of θ in respect of point L' for the extremities M_1, M_2 of the curve M_1AM_2, would be

$$\frac{\lambda ii'}{2g}(\cos \theta_1' - \cos \theta_2').$$

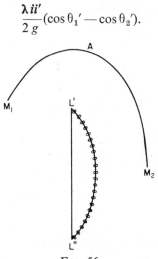

FIG. 56.

Consider now the definite solenoid $L'L''$ (Fig. 56) which may only revolve about the axis through its two extremities. We shall again be able to replace it by two indefinite solenoids; the sum of the actions of the current M_1AM_2 on each of them is equivalent to its action on $L'L''$. The rotational moment of the first has just been found; putting θ_1'', θ_2'' for the angles corresponding to θ_1', θ_2', but in respect of the extremity L'', that of the second is

$$-\frac{\lambda ii'}{2g}(\cos \theta_1'' - \cos \theta_2'');$$

the total moment produced by the action of M_1AM_2 for rotation of the solenoid about its axis $L'L''$ therefore is

$$\frac{\lambda\,ii'}{2\,g}\,(\cos\theta_1' - \cos\theta_1'' - \cos\theta_2' + \cos\theta_2'').$$

This torque is independent of the shape of the conductor M_1AM_2, its magnitude and its distance from the solenoid $L'L''$, and it remains so as long as any such variation entails no change in the angles θ_1', θ_2', θ_1'', θ_2''; it is zero not only when the current M_1M_2 forms a closed circuit, but also when the current is assumed to extend to infinity in both directions, because in that event, the two extremities of the current being infinitely remote from the extremities of the solenoid, the angle θ_1' becomes equal to θ_1'', and θ_2', and θ_2' equals θ_2''.

All the moments of rotation about straight lines drawn through the extremity of an indefinite solenoid being zero, this extremity is the point at which the resultant of the forces exerted on the solenoid is applied by a closed circuit, or by a system of currents forming more than one closed circuit; it may therefore be assumed that all these forces are transported there and this point may be taken as

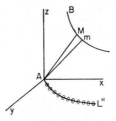

FIG. 57.

the origin of coordinates (Fig. 57): suppose that BM is a portion of one of the currents acting on the solenoid. From the foregoing the force due to some element Mm of BM is normal to the plane AMm and represented as

$$\frac{\lambda\,ii'\,d\nu}{g\,r^3},$$

where $d\nu$ is the area AMm and r is the variable distance AM.

To obtain the component of this action along AX, it has to be multiplied by the cosine of the angle which it forms with AX, which is the same as the angle between the planes AMm, ZAY; but $d\nu$, multiplied by this cosine, is the projection of AMm on ZAY, which is equal to

$$\frac{y\mathrm{d}z - z\mathrm{d}y}{2} :$$

If therefore it is desired to find the action exerted along AX by currents forming closed circuits, it is necessary to take the following integral over the entire range of the currents

$$\frac{\lambda\,i\,i'}{2g} \int \frac{y\mathrm{d}z - z\mathrm{d}y}{r^3} \quad \text{which is} \quad \frac{\lambda\,i\,i'\,A}{2g},$$

the quantity A being the same as before, where n was replaced by its value 3; likewise the action along AY is

$$\frac{\lambda\,i\,i'\,B}{2\,g},$$

and along AZ

$$\frac{\lambda\,i\,i'\,C}{2\,g}.$$

The resultant of these three forces, which is the total action exerted by a number of closed circuits on an indefinite solenoid, is therefore equal to

$$\frac{\lambda\,i\,i'\,D}{2\,g},$$

where $D= \sqrt{(A^2 + B^2 + C^2)}$; and the cosines of the angles which it forms with the axes of x, of y and of z are:

$$\frac{A}{D}, \quad \frac{B}{D}, \quad \frac{C}{D},$$

which are the values of the cosines of the angles between the same axes and the normal to the directing plane as if the action of the circuits on an element situated at A were considered. Now this element would be transported by the action of the system in a direction contained within the directing plane; hence the remarkable conclusion is reached that when a system of closed circuits acts alternately on an indefinite solenoid and on an element situated at the extremity of this solenoid, the respective directions in which the element and the extremity of the solenoid are carried, are mutually perpendicular. If the element is itself situated in the directing plane, the action exerted upon it by the system is at its maximum and equal to

$$\frac{i\,i'\,D\,ds'}{2}.$$

The action which this system exerts on the solenoid was found just now to be

$$\frac{\lambda\,i\,i'\,D}{2\,g}.$$

These two forces are therefore always in a constant ratio for a particular element and a particular solenoid which is equal to

$$ds' : \frac{\lambda}{g}.$$

That is to say, the forces are in the same relation as the length of the element bears to the area of the closed curve described by one of the currents of the solenoid divided by the distance between two consecutive currents; this ratio is independent of the form and magnitude of the currents of the system acting on the element and solenoid.

If the system of closed circuits is itself an indefinite solenoid, the normal to the directing plane through point A is, as we have just seen, a straight line joining point A to the extremity of the solenoid; hence the mutual action of two indefinite solenoids takes place among the straight line joining the extremity of one solenoid to the extremity of the other; in order to determine its value, we put λ' for the area of the circuits formed by the currents of this new solenoid, g' for the distance between the planes of two of these consecutive circuits, l for the distance between the extremities of the two indefinite solenoids, and we get $D = -\ \lambda'/g'l'$, which yields for their interaction,

$$\frac{\lambda\,i\,i'\,D}{2\,g} = -\ \frac{\lambda\lambda'\,i\,i'}{2\,g\,g'\,l^2},$$

which is in inverse ratio to the square of the distance *l*. When one of the solenoids is definite, it can be replaced by two indefinite solenoids and the action is then made up of two forces, one attractive and the other repellent, along the straight lines which join the two extremities of the first solenoid to the extremity of the other. Finally,

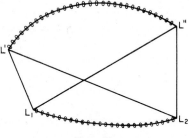

FIG. 58

if two definite solenoids $L'L''$ and L, L interact (Fig. 58), there are four forces along the respective straight lines $L'L_1$, $L'L_2$, $L''L_1$, $L''L_2$ which join the extremities in pairs; and if, for example, there is repulsion along $L'L_1$, there will be attraction along $L'L_2$ and $L''L_1$, and repulsion along $L''L_2$.

In order to justify the manner in which I have conceived magnetic phenomena, regarding magnets as assemblies of electric currents forming minute circuits round their particles, it should be shown from consideration of the formula by which I have represented the interaction of two elements of current, that certain assemblies of little circuits result in forces which depend solely on the situation of two determinate points of this system. These are endowed with all the properties of the forces which may be attributed to what are called molecules of austral fluid and of boreal fluid, whenever these two fluids are used to explain magnetic phenomena, whether in the mutual action of magnets, or in the action of a magnet on a con-ductor. Now the physicists who prefer explanations based on the existence of such molecules to the explanation which I have deduced from the properties of electric currents, are known to admit that each molecule of austral fluid always has a corresponding molecule of boreal fluid of the same intensity in each particle of the magnetized body. In saying that the assembly of these two molecules, which may be regarded as the two poles of the element, is a magnetic element, an explanation of the phenomena associated with the two kinds of action in question requires: (1) that the mutual action of magnetic elements should be made up of four forces, two attractive and two repellent, acting along straight lines joining the two molecules of one of these elements to the two molecules of the other, with intensity in inverse ratio to the squares of these lines; (2) that when one of these elements acts on an infinitesimal portion of conducting wire, two forces result, perpendicular to the planes passing through the two molecules of the element and the small portion of wire, and proportional to the sines of the angles between the wire and the straight lines joining the wire to the two molecules, and which are in inverse ratio to the squares of these distances. So long as my concept of the behaviour of a magnet is disputed and so long as the two types of force are attributed to molecules of austral and boreal fluid, it will be impossible to reduce them to a single principle; yet no sooner than my way of looking at the constitution of magnets is adopted, it is seen from the foregoing calculations that the actions of these two kinds and the values of the resulting forces are deducible directly from my formula. To determine their values it is sufficient to replace the assembly of two molecules, the one of austral and the other of boreal fluid, by a solenoid with extremities that are the two determinate points on which the forces in question depend, and which are situated at precisely the same points where it is assumed that the molecules of the two fluids are placed.

Two systems of very small solenoids then act on each other, according to my formula, like two magnets composed of as many magnetic elements as there are assumed to be solenoids in the two systems. One of these systems will also act on an element of electric current in the same way as a magnet.† In consequence, in as much as

---

† I believe I must here introduce the following note which is taken from the analysis of the works of the Académie in the year 1821, published 8 April 1822.

"The principal difference between the manner in which a magnet acts and the action of a conductor of which one part is rolled in a helix round the other, is that the poles of the magnet are situated nearer to the mid-point of the magnet than its extremities, whereas the points which present the same properties in the helix occur exactly at the extremities of this helix: this is what must take place as the intensity of the currents

all calculations and explanations are based either on the attractive and repellent forces of the molecules in inverse ratio to the squares of the distances, or on the rotational forces between a molecule and an element of electric current the law governing which I have just indicated as accepted by physicists who do not accept my theory, they are necessarily the same whether the magnetic phenomena in these two cases is explained in my way by electric currents, or whether the hypothesis of two fluids is preferred. Objections to my theory, or proofs in its favour, therefore, are not to be found in such calculations or explanations. The demonstration on which I rely results all from the fact that my theory explains in a single principle three sorts of actions that all the associated phenomena proves are due to one common cause. This cannot be done otherwise. In Sweden, Germany and England it has been thought possible to explain the phenomena by the interaction of two magnets as determined by Coulomb. Experiments which produce continuous rotational motion are manifestly at variance with this idea. In France, those who have not adopted my theory, are obliged to regard the three kinds of action which I have interrelated, as though absolutely independent. The law which Coulomb established in respect of the action of two magnets could be deduced from the law proposed by M. Biot for the mutual action of a portion of conducting wire and a "magnetic molecule"; but if it is admitted that one of

---

of the magnet diminishes outwards from the mid-point to its extremities. But M. Ampère has since recognized that another factor can also cause this effect. Having concluded from his more recent experiments that the electrical currents of a magnet exist round each one of its particles, it was easier for him to see that it is unnecessary to assume, as he had done at first, that the planes of these currents are everywhere perpendicular to the axis of the magnet. Their interaction, however, must tend to incline the planes to the axis, particularly towards the extremities, so that the poles, instead of being exactly situated there, as they ought to be according to the calculations by the formulae of M. Ampère for the case when all the currents are assumed to be the same in intensity and in planes perpendicular to the axis, must be closer to the mid-point of the magnet according as the planes of a larger number of currents are thus inclined, and yet more so with greater inclination. That is to say, the thicker the magnet in relation to its length the greater is this effect, which agrees with experiments. In helical conductors, where one part returns along the axis to cancel the effect of the part of the currents of each turn which acts as

these magnets is composed of small electric currents, like those which I have suggested, how can it be objected that the other is not likewise composed, thereby accepting all of my view?

Moreover, though M. Biot determined the value and direction of the force when an element of conducting wire acts on each particle of a magnet and defined this as the elementary force*, it is clear that a force cannot be regarded as truly elementary which manifests itself in the action of two elements which are not of the same nature, or which does not act along the straight line which joins the two points between which it is exerted. In the mémoire which this gifted physicist communicated to the Académie the 30 October and 18 December 1820‡, he still regarded the force which an element of conducting wire exerts on a molecule of austral or boreal fluid as elementary, that is to say, the action exerted on the pole of a magnetic element is regarded as elementary.

When M. Oersted discovered the action which a conductor exerts on a magnet, it really ought to have been suspected that there could be interaction between two conductors; but this was in no way a necessary corollary of the discovery of this famous physicist. A

---

though they were parallel to the axis, the two circumstances which, from the foregoing remarks, do not necessarily occur in magnets, must on the contrary exist in the conductors; it may also be observed in experiments that helices have poles similar to those of magnets, but placed exactly at their extremities as calculated."

It will be seen from this note that after 1821 I concluded from the phenomena associated with magnets: (1) that in regarding each particle of a bar magnet as a magnet, the axes of these elementary magnets must be, not parallel to the axis of the total magnet as was supposed at the time, but inclined to this axis in directions determined by their interaction; (2) that this is one reason why the poles of a bar magnet are not at the extremities, but in between the extremities and its mid-point. Both these conclusions have already been demonstrated by the results of M. Poisson using the formulae by which he represented the distribution of the forces emanating from each particle of a magnet. These formulae are based on Coulomb's law and, in consequence, nothing is changed by adopting my approach to the explanation of magnetic phenomena, since this law is a corollary of my formula, as has been seen in the course of this paper.

*Précis élémentaire de physique, vol. II, p. 122, 2nd edit.

‡Since the latter mémoire has not been published separately, the formula for the force is only known to me from the following passage in the second

bar of soft iron acts on a magnetized needle, but there is no inter-action between two bars of soft iron. Inasmuch as it was only known that a conductor deflects a magnetized needle, could it have been concluded that electric current imparts to wire the property to be influenced by a needle in the same way as soft iron is so influenced without requiring interaction between two conductors when they are beyond the influence of a magnetized body? Only experiments could decide the question; I performed these in the month of September 1820, and the mutual action of voltaic conductors was demonstrated.

It was of little value that I should merely have discovered the action of the earth on a conductor and the interaction of two conductors and verified them by experiments; it was more important:

(1) To find the formula for the interaction of two elements of current.

(2) To show by virtue of the law thus formulated (which governs the attraction of currents in the same direction and the repulsion of currents in the opposite direction, whether the currents are parallel or at an angle), that the action of the earth on conducting wires is identical in all respects, to the action which would be exerted on the same wires by a system, (*fasces*, Latin) of electric currents flowing in the east–west direction, when situated in the middle of Europe where the experiments which confirm this action were performed.

(3) To calculate first, from consideration of my formula and the manner in which I have explained magnetic phenomena associated with electric currents forming very small closed circuits round particles of a magnetized body, the interaction between two particles

---

edition of *Précis élementaire de physique*, vol. II, pp. 122–3.
"By imagining the length of the connecting wire $Z'C'$ (Fig. 34) to be divided into infinitely many very fine sections, it is seen that each section must act on the needle with a different energy according to its distance and direction. Now, these elementary forces are just the simple result which it is especially important to know; for the total force exerted by the complete wire is nothing other than the sum of their individual actions. However, calculation is sufficient to analyse from the resultant the simple action. This is what Laplace did. He deduced from our observations that the individual law of the elementary forces exerted by each section of the connecting wire was in the inverse ratio of the square of the distance, that is to say, it is precisely the same as what is known to exist in ordinary

of magnets regarded as two little solenoids each equivalent to two magnetic molecules, the one of austral and the other of boreal fluid, and the action which one of these particles exerts on an element of conducting wire; then to check that these calculations give exactly, in the first case the law established by Coulomb for the action of two magnets, and in the second case, the law which M. Biot has proposed for the forces which develop between a magnet and a conducting wire. It is thus that I reduced both kinds of action to a single principle and also that which I discovered exists between two conducting wires. Doubtless it was simple, having assembled all the facts, to conjecture that these three kinds of action depended on a single cause. But it was only by calculation that this conjecture could be substantiated, and this is what I have done. I draw no premature conclusion as to the nature of the force which two elements of conducting wires exert on each other, for I have sought only to obtain the analytical expression of this force from experimental data. By taking this as my starting point I have demonstrated that the values of the other two forces given by the experiment (the one between an element of conducting wire and what is called a magnetic molecule, the other between two of these molecules) can be deduced purely mathematically by replacing, in one or the other case, as is necessary, according to my conception of the constitution of magnets, each magnetic molecule by one of the two extremities of an electro-dynamic solenoid. Thereafter, all that can be deduced from these values of the forces is necessarily contained in my manner of considering the effects which are produced and it becomes a corrollary of my formula, and that alone should be sufficient to demonstrate

---

magnetic actions. The analysis showed that to complete our knowledge of the force, it remained to determine whether the action of each section of the force was the same in all directions at the same distance, or whether the energy was greater in some directions than in others. To decide this question, in the vertical plane I bent a long copper wire ZMC at M (Fig. 34) in such a way that the two arms ZM, MC were at the same angle of the horizontal MH. In front of this wire I stretched another piece $Z'M'C'$ of the same material, the same in diameter and of the same grade; this piece I set up vertically, being separated from the first piece at MM only by a strip of very fine paper. I then suspended the magnetized needle AB in front of this system at the height of the points M, M' and observed the oscillations at various distances whilst passing current successively

that the interaction of two conductors is, in fact, the simplest case and that from which it is necessary to proceed in order to explain all other cases. The following considerations seem to finish a complete confirmation of these general results of my work; they are founded on the simplest of notions about the composition of forces in reference to the interaction of two systems of infinitely close points in the various cases which can arise—whether these systems only contain points of the same type, that is to say, points which attract or repel similar points of the other system, or whether one of the systems, or both, contains points of the two opposite types of which those of one type attract what those of the other repel, and repel what they attract.

Throughout history, whenever hitherto unrelated phenomena have been reduced to a single principle, a period has followed in which many new facts have been discovered, because a new approach in the conception of causes suggests a multitude of new experiments and explanations. It is thus that Volta's demonstration of the identity of galvanism and electricity was accompanied by the construction of the electric battery with all the discoveries which have sprung from this admirable device. Judging from the important results of the work of M. Becquerel on the influence of electricity in chemical compounds, and that of MM. Prévost and Dumas on the causes of muscular contraction, it may be hoped that their discovery of new knowledge over the past four years and its reduction to a single principle of the laws of attractive and repellent

---

through the bent and straight wires. In this way I found that the action was reciprocal for both wires to the distance to the points M, M'; but the absolute intensity was weaker for the oblique wire than for the straight wire in the same proportion that the angle ZMH is to unity. An analysis of this result appears to indicate that the action of each element μ of the oblique wire on each molecule m of austral or boreal magnetism is reciprocal to the square of its distance μm to this molecule and proportional to the sine of the angle $m\mu M$ beween the distance μm and length of the wire."

It is remarkable that this law, which is a corollary of the formula by which I have represented the interaction of two elements of conducting wires when, according to my theory, each magnetic element is replaced by a very small electrodynamic solenoid, was first found through a mathematical error; indeed, for the law to be valid, the *absolute intensity* ought to have been proportional, not to the angle ZMH, but to the tangent of

forces between electric conductors, will also lead to a host of other results which will establish the links between physics, on the one hand, and chemistry and even physiology, on the other, for which there has been a long-felt need, though we cannot flatter ourselves for having taken so long to realize it.

It still remains to consider the actions exerted by a closed circuit of arbitrary shape, magnitude and position on a solenoid, or on some other circuit of arbitrary shape, magnitude and position; the principal result from such inquiries is the similarity which exists between the forces produced by a circuit, whether acting on another closed circuit or a solenoid, and the forces which would have been exerted by points whose action were precisely that which is attributed to molecules of what is called austral and boreal fluid. Let us assume that these points are distributed in the manner which I have just explained over surfaces terminated by circuits, and that the extremities of the solenoid are replaced by two magnetic molecules of opposite types. The analogy seems at first to be so complete that all electrodynamic phenomena appear to be reduced to the theory associated with these two fluids. It is soon seen, however, that this only applies to conductors which form solid and closed circuits, that it is only phenomena which are produced by conductors forming such circuits that may be explained in this way, and that in the end it is only the forces which my formula represents that fit all the facts.†

---

half this angle, as demonstrated later by M. Savary in his dissertation at the Académie, 3 February 1823, and which has meanwhile been published in the *Journal de physique*, vol. XCVI, pp. 1–25 cont'd. It appears that M. Biot later discovered the error himself, for in the third edition which has just appeared, he describes, without reference to the Mémoire where it had first been corrected, new experiments where the intensity of the total force is, in accordance with the calculation of M. Savary, proportional to half the angle ZMH, and he concludes therefrom, with more reason than he had with his first experiments, that the force which he calls elementary, is proportional for equal distance to the sine of the angle between the direction of the element of conducting wire and the direction of the straight line joining its mid-point to the magnetic molecule. (*Précis élementaire de physique expérimentale*, 3rd edit., vol. II, pp. 740–5).

† It seems at first that the effects produced by magnets, or by sets of solid and closed circuits, ought only to be identical for closed circuits of

very small diameter; but it may readily be seen that it is also true of circuits of arbitrary magnitude since, as we have seen, they may be replaced by magnetic elements distributed uniformly over surfaces terminated by these circuits, whilst the number of surfaces that a particular circuit circumscribes can be multiplied as you please. The set of surfaces may be regarded as a 'system' of magnets which are equivalent to the circuit. The same consideration proves that without in any way affecting the resulting forces, the infinitesimal currents which encircle the particles of a bar magnet can always be replaced by currents of finite dimension, these currents forming closed circuits about the axis of the bar when those of the particles are distributed symmetrically about this axis. For this it is sufficient to imagine surfaces within the bar terminating at the surface of the magnet and cutting the lines of magnetization everywhere at right angles and passing through the magnetic elements which can always be assumed to be placed at the points where these lines are met by the surfaces. Then, if all the elements of a particular surface are of equal intensity on equal areas, they can be replaced by a single current flowing along the curve formed by the intersection of this surface and that of the magnet. If they should vary, increasing in intensity from the surface to the axis of the magnet, they would first be replaced by a current at this intersection such as it ought to be according to the *minimum* intensity of the particular currents of the surface normal to the lines of magnetization under consideration, and then, for each line circumscribing the portions of this surface where the little currents become more intense, a new current should be imagined which is concentric to the previous one as required by the difference in intensity of the adjacent currents, some outward and the others inward of this line. If the intensity of the particular currents decreases from the surface to the axis of the bar, a corresponding concentric current should be imagined on the separation line in the opposite sense. Finally, an increase of intensity which might follow the decrease would require a new concentric current directed as in the first case.

These comments are only given here so as not to omit a remarkable conclusion which may be drawn from the results of this paper; they are in no way intended to corroborate the supposition that the electric currents of magnets form closed circuits about their axes. Having at first hesitated between this supposition and the other way of regarding currents as encircling the particles of magnets, I have recognized for a long time that this latter concept best fitted all the facts and in this respect my opinion has not changed at all.

Moreover, this conclusion is useful in that it identifies the actions produced by an electrodynamic helix, on the one hand, or by a magnet, on the other, just as completely from the point of view of theory as when verified by experiments.

5
Grassmann

A NEW THEORY OF ELECTRODYNAMICS*

It is well known that the dynamic effects exerted by electric
currents or magnets on other electric currents or magnets, as far as
our observations have gone, may be explained on the basis of a
single principle. But the extent of these observations, as I shall show,
leaves room for discussion as to the basis on which the mutual
interaction of two portions of a current is to be explained. When I
submitted the explanation offered by Ampère for the interaction of
two infinitely small current-sections on one another to a more
exacting analysis, this explanation seemed to me a highly improbable
one; and when I then tried to eliminate the arbitrary element in this
explanation, another explanation occurred to me which was able to
elucidate electrodynamic phenomena (in so far as they have at
present been observed) with the same exactitude, and which seemed
particularly likely to be correct in view of the simplicity of the
fundamental formulae and of the complete similarity which it
showed to all other dynamic forces. I have already indicated that this
new explanation, when applied to all phenomena observed up to
now, gives the same results as that of Ampère; but there exists a
range of phenomena, on the other hand, for which the two explanations give diametrically opposed results: it is therefore these phenomena which must constitute the decisive ones as to which of the two
explanations is to be regarded as correct. The field in which such
phenomena lie is that in which opposite electric charges are imposed
(as by an electric machine) at the ends of a conductor, and so produce
a current-flow. Experiments hitherto made in this field, in which the
dynamic effects were expected to reveal themselves by the deflection

*Poggendorff's *Annalen der Physik und Chemie*, **64** (1), 1–18 (1845) of a magnetic needle, for example, are entirely inadequate to reveal the difference between the two hypotheses; while other experiments which might be made for this purpose have up to now been confronted with serious difficulties. It seems to me, however, important to indicate the predictions which the two explanations offer, so that finer instruments and more accurate observations may subsequently indicate which is to be regarded as the more probable. To do this would give a guide whereby experienced workers could perhaps devize decisive experiments. It will therefore be my task to derive the new explanation, and that of experienced physicists to test it experimentally.

(1) All experiments hitherto set up to examine electrodynamic phenomena have either been conducted with closed circuits, or with circuits which may be regarded as though they were closed.† These experiments consist in observing either the mutual interaction of two closed circuits, or by making one part of a closed circuit movable, observing both the interaction which it experiences from the entire circuit to which it belongs, and also the alteration of this interaction which occurs when other closed circuits are introduced. Since to make one part of a circuit movable causes no alteration in the effect produced by the whole circuit, experiments hitherto made only extend to the effect of closed circuits on other closed circuits or on circuit elements. On the other hand, no experiments have been set up to test the effect of a part of a circuit, either upon a closed circuit or upon another circuit portion.

(2) Ampère was obliged, therefore, in order to obtain his formula, to use an arbitrary assumption together with the experimental results. The assumption used for this purpose is, at first glance, very simple and natural, consisting in the supposition that two infinitely small circuit elements exert forces on each other along the straight line connecting their mid-points, either of attraction or repulsion. By means of this assumption Ampère is able to proceed

---

†Amongst these are the deflection of the magnetic needle by the discharge of a battery in which, on the one hand, the numerous windings of the multiplier and, on the other, the proximity of equalized electric charges separated only by the thickness of the glass, cause the currents in the circuits, as measured by their observable effects to be the same as though the circuits were closed.

from the experimental results directly to his basic formula, according to which the force exerted by one infinitely small circuit element a on another such element b is proportional to the expression:

$$(2 \cos \varepsilon - 3 \cos \alpha . \cos \beta) . ab/r^2, \qquad (1)$$

where a and b are the circuit elements, that is, the infinitely small linear portions multiplied by the current intensity, in which the currents move; r is the distance of the mid-points of these elements from each other; ε is the angle between the two circuit portions; and α and β are the angles formed by the elements a and b respectively with the line drawn between the two mid-points.

(3) The complicated form of this formula arouses suspicion, and the suspicion is heightened when an attempt is made to apply it. If, for example, the simplest case is considered, in which the circuit elements are parallel, so that $\varepsilon = 0$ and $\alpha = \beta$, the Ampère expression becomes

$$(2 - 3 \cos^2 \alpha) . ab/r^2,$$

from which it appears that, when $\cos^2 \alpha$ is equal to $\frac{2}{3}$ or, which comes to the same thing, $\cos 2\alpha$ is equal to $\frac{1}{3}$, that is if the position of the mid-point of the attracted element lies on the surface of a cone whose apex is at the attracting element, and whose apex angle is arccos $\frac{1}{3}$, there is no interaction; while for smaller angles there is repulsion, and for larger ones attraction. This is such an unlikely result, that the principle from which it is derived must come under the gravest suspicion and with it the supposition that the force in question must show an analogy with all other forces. It must be concluded that there is little reason to apply this analogy to our present field. Since in the case of all other forces it is originally point elements, without any definite direction, which interact with each other, so that the mutual interaction must *a priori* be regarded as necessarily operating along the line connecting them, it is hard to see any justification for transferring this analogy to an entirely foreign field in which the elements are arranged in definite directions. The formula itself, which in no way resembles that for gravitational attraction, also indicates that there is no real analogy.

(4) Without making any arbitrary assumption of my own, therefore, I propose to eliminate the arbitrary factor in the Ampère

hypothesis, while recognizing, as I must, that the Ampère hypothesis has been completely vindicated in such experimental tests as
have hitherto been applied to it, which concern attractions existing
between closed circuits on other circuits or circuit elements. It is
easily seen that all phenomena lying within such a field may be
explained, if the interaction exercised by an angular current—
that is, an infinite current flowing along the arms of an angle—
on a circuit element whose mid-point lies in the plane of the angle, is
known. Since, in the first place, I may consider any closed or open
circuit as made up of such circuit elements, and, in the second place,
I may regard any closed circuit as made up of a polygon traversed
by the current, this polygon itself consisting of angular currents
forming the outer angles of the polygon, relying upon experience
for the statement that equal and opposite currents which flow
through the same conductor successively annul one another. Thus
the circuit *abc* (Fig. 59) may be regarded as composed of the three
angular circuits *fad, dbe, ecf*. Finally, if I construct a line between
the mid-point of the attracted element and the apex of the angular
circuit, I may break down the latter into two angular circuits, each
of which lies in the same plane with the mid-point of the attracted
element. Thus, in order to remove the arbitrary element from the

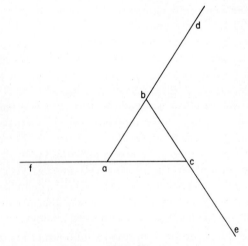

FIG. 59

Ampère concept it is only necessary to derive the effect of an angular circuit on an element lying in the same plane.

(5)   It follows at once from the Ampère formula that the inter-action experienced by one element from another, if the two elements do not lie in the same plane, is equal to the interaction which the perpendicular projection of the former on to the plane drawn through its mid-point and containing the latter element experiences. This relation will thus also apply to our present case, so that it is only necessary to look for the effect of an angular circuit on an ele-ment in the same plane, or in a line drawn through such an element and lying next to it. This action can be broken down into one lying along the element, and another perpendicular to it.

(6)   For longitudinal movement it appears that this is independent of the direction of the line,† and thus for an angular current is as large as if both lines coincided: that is, it is equal to zero. It follows from this that the effect produced by an angular circuit on an element lying in its own plane must be perpendicular to the latter, so that, in passing, it follows that the effect of any closed circuit on a circuit element is always perpendicular to the latter.

---

† Since from Ampère's formula, if ds is an element of the line (considered in the direction of the line), and i is the current intensity proceeding from the starting point of this line, the attraction exerted by this element on the circuit element b along its length is given by the expression:

$$- (ids.b/r^2) \cos \beta\ (2 \cos \varepsilon - 3 \cos \alpha . \cos \beta).$$

Further, if l is the perpendicular from the mid-point of the attracted ele-ment on to the line of the attracting one (see Fig. 60), ds is given by

$$sds = - d(l \cot \alpha) = l.d\alpha/\sin^2\alpha = r^2.d\alpha/l.$$

where $\varepsilon = \alpha - \beta$, and $d\beta = d\alpha$. Hence we obtain the expression

$$- (ib/l)\ (\cos^2\beta . \cos \alpha.d\alpha - 2 \sin \alpha . \sin \beta . \cos \beta.d\beta)$$

which gives on integration:

$$- (ib/l) \sin \alpha . \cos^2\beta.$$

If the integration is taken over the entire line, and finally the values of α, β and r are taken as those corresponding to the starting point of the line, then, replacing l by its value $r.\sin \alpha$, we obtain the expression

$$(ib/r) \cos^2\beta$$

for the longitudinal effect of the line, which is independent of α, and thus of the direction of the line.

(7) The force perpendicular to the attracted element, which according to the formula of Ampère is exerted on such an element by a current-carrying circuit lying in the same plane, appears in the

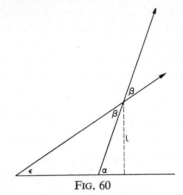

FIG. 60

form of two terms, of which one is independent of the direction of the attracting circuit-line, and therefore vanishes on the angular-current assumption, while the other is given by:[†]

$$(ib_l/r) \cot \tfrac{1}{2} \alpha \qquad\qquad (2)$$

where r is the distance of the element from the initial point of the line,

---

[†]That is, maintaining the expression above, the effect of an element $i.ds$ of the line through which current flows on the current element b_l in a perpendicular direction to the latter is equal to

$$(i.ds.b_l/r^2) \sin \beta \ (2 \cos \varepsilon - 3 \cos \alpha . \cos \beta),$$

which then, since $ds/r^2 = d\alpha/l$, $d\alpha = d\beta$, $\varepsilon = \alpha - \beta$, so that $2 \cos \varepsilon = \cos \varepsilon + \cos (\alpha - \beta)$,

$$(ib_l/l) (\cos \varepsilon . \sin \beta . d\beta - 2 \cos \alpha . \sin \beta . \cos \beta . d\beta + \sin^2\beta . \sin \alpha . d\alpha),$$

and thus gives on integration:

$$- (ib_l/l) (\cos \varepsilon . \cos \beta + \cos \alpha . \sin {}^2\beta),$$

which, if the integration is carried out over the entire line, and the designation of the variable quantities (α, β) is limited to their values applying to the initial point of the line, gives the expression

$$(ib_l/l) (1 + \cos \varepsilon . \cos \beta + \cos \alpha . \sin {}^2\beta),$$

in which at infinity α becomes $180°$, and β becomes $180° - \varepsilon$. If, finally, l and ε are replaced by their values $r \sin \alpha$ and $(\alpha - \beta)$, the term $\cos \alpha . \cos^2\beta$ combines with the term $\cos \alpha . \sin^2\beta$ to give the single term $\cos \alpha$, and if $(1 + \cos \alpha)/\sin \alpha$ is then replaced by $\cot \tfrac{1}{2}\alpha$, we obtain

$$(ib_l/l) (\cot \tfrac{1}{2}\alpha + \sin \beta . \cos \beta)$$

of which the second term is independent of α, and so of the direction of the attracting ray.

and α is the angle between the line and that drawn from its starting point through the attracted element; while b_l is the perpendicular projection of the element on the plane through its centre and the current-line, while i is the intensity of the current passing through the line. Finally, the current element moves to its right or left according to whether the current flows to the right or the left considered from the element. From this the effect of an angular current, the arms of which make angles α and α' with the line drawn through the attracted element, is given by:

$$(ib_l/r) \cdot (\cot \tfrac{1}{2}\alpha - \cot \tfrac{1}{2}\alpha') \tag{3}$$

It follows from this, incidentally, that the magnitude of the force experienced by a current element from a current lying in the same plane, is independent of the direction of this element, but is always perpendicular to this and towards the same side.

(8) Expression (3) for the effect of an angular current is now entirely free from any hypothetical element, since the effect can at least approximately be obtained by means of experiment, while it also contains the experimental results, since these can all be traced to the effect of angular currents, and may be therefore used as the foundation for any hypothesis on the mutual interaction of current elements. Since this expression consists of two terms, of which one is determined by the position of the one ray, and the other by that of the second, it is clearly simplest to take these terms as expressions for the effects of the individual rays; that is, expression (2) is taken as the actual expression of the attraction of a ray traversed by a current: in fact any other assumption introduces irrelevant material into the formula, and so appears artificial. I therefore regard expression (2), that is, $(ib \cot \tfrac{1}{2}\alpha/r)$, as the expression of the effect of a ray in the sense given above, and use this as the basis of the following development.

(9) From this we at once obtain the mutual interaction of two current elements, since we may regard the attracting current element $i.ds$ as the combination of two lines through which current is passing, these possessing the direction and intensity (i) of this element, and one of them having its current flowing in the same direction as that in the element, and the other in the opposite direction, while the first of them has its starting point in the initial point of the element, and

the second has its starting point in the end point of the element. We then obtain

$$(ab_l/r^2) \cdot \sin \alpha \qquad (4)$$

as the expression for the effect exerted by a current element a on another b, distant r from it, the vertical projection of the second element on the plane through a and r being equal to b_l, while α represents the angle formed between a and the line drawn to b. The movement then occurs perpendicularly to b (or b_l) in the plane through a and r, towards that side to which the side a of the angle α lies with respect to the other side of the angle (see Fig. 60).†

(10)   We consider first the mutual interactions of two current elements a and b whose prolongations intersect: in this case it is clear that both movements, since they are perpendicular to the moving current elements, can be regarded as brought about by rotation of the two straight lines to which the current elements belong. Then the angle through which one of the lines (say that to which element b belongs) rotates is given by the movement of the element divided by the distance (B) of this element from the intersection. It is thus given by‡:

$$(ab \sin \alpha)/r^2 B = (ab \sin \varepsilon)/r^3 \qquad (5)$$

From this formula it is seen that the lines in which the two elements lie pass through the same angle during the movement, while their intersection does not move, and the position of the elements in the lines is also unchanged. It is also easily seen that the angle between the lines is reduced during movement if the current in both elements moves either towards or away from the point of intersection, while the angle is increased if one current moves towards the intersection while the other moves away from it. Thus the truly reciprocal nature of the movement becomes apparent, and it is seen that this reciprocal attraction of two linear portions has just the same effect in reducing the angle between them for a constant point of intersection,

---

†It is then only necessary to differentiate expression (2) with respect to —ds in order to find the attraction of the element ids. Instead of r, cot $\frac{1}{2}\alpha$ and dα, we then employ their values $l/\sin \alpha$, $(1 + \cos \alpha)/\sin \alpha$ and lds/r^2, and then, by differentiation, the desired expression

$$(ib_l.ds/r^2) \sin \alpha \quad \text{or} \quad (ab_l/r^2) \sin \alpha.$$

‡Since $(l/B) \sin \alpha = (l/r) \sin \varepsilon$. See Fig. 60.

as the reciprocal attraction between two points has in reducing the distance between them when they continue to lie on the same line. We thus obtain, instead of the artificial and superficial analogy involved in the Ampère assumption, a true and natural similarity, since here lines and points in a plane correspond to one another in the same way as angle and distance, and as point of intersection and connecting line.

(11)   This analogy becomes even clearer when it is shown that the electrodynamic attraction according to the new theory is expressible by the same formula as the attraction due to gravity. For this purpose it is here necessary to introduce the idea of a connection which I have indicated in a work recently issued,† worked out before I had any idea of the theory at present under discussion. I have there shown that the product of two points a and b must take account of their connecting distance, so that the product of two points loaded with given intensities (or weights) must, in addition to the product of the intensities, contain a term characterizing the separation. Thus, if α and β are points, the distance $\alpha\beta$ between which denotes not merely the magnitude of the separation but also its direction, and if the intensities at the points are taken as 2 and 3 respectively, the value of a must be taken as 2α and that of b to be 3β, in order to attain the product ab. I have shown in the same place that this product differs from the arithmetical system in that ab is equivalent to $-ba$. According to this the distance must be multiplied by 6 to maintain the value ab, and so the attraction exercised by a point a on another point b distance r from it, due to gravity, is for any value of the weights proportional to

$$ab/r^3 \qquad (6)$$

an expression which includes the direction as well as the intensity of the attraction. I have also shown in the same place that the area of a parallelogram may be regarded as the product of two adjacent sides a and b if both the length and direction of these sides is taken into account, and that here also $a.b$ will be equal to $-ba$. I have also shown that if the lines in which a and b lie are related to these magnitudes, then the product represents the collected point of

---

†*Ausdehnungslehre* (*Dimensional Theory*). Part 1, containing linear dimensional theory. Refer to pp. 61, 164 and 222.

intersection of the two lines for each surface area. Now the numerator of expression (5) is clearly the expression for the area of a parallelogram which has a and b as its sides in their appropriate directions. Hence, if a and b are taken to be the current elements, represented by the lines along which they lie, expression (5) is converted to (6):

$$ab/r^3,$$

which is identical with that arrived at for the attraction due to gravity, and whose magnitude expresses that of the deflection which the two elements try to produce, while the point represented at the same time by the product ab gives the centre of deflection.

(12)   We have only demonstrated the analogy when the current elements intersect on prolongation. But the case for which the current elements are parallel is not essentially different, since this can be considered in such a way as to suppose that they would intersect at infinite distance. On the other hand, the situation is more difficult if the current elements are not found in the same plane. For such a case I will only note that the movement may be resolved into two movements of the lines belonging to each element, using the mutual perpendicular of the two lines (their shortest distance apart) in place of the point of intersection. One of these movements consists of a deflection about this common perpendicular, which then causes a reduction or an increase of the angle between the two currents; while the other movement consists in a reduction or an increase of the perpendicular, which operates in such a way that these lines move towards the perpendicular. In both cases the movement is a mutual one, and the lines remain perpendicular to the mutual perpendicular, and the current elements do not alter their position within this line. It is easily seen that we have here the most complete analogy with the movement caused by gravity. I could also show that this movement, also, is capable of representation by means of expression (6). It is not possible, however, to develop the evidence for this without developing the principles of an analysis which is of the greatest importance to physics (and can often represent the most apparently complex relations in the simplest form); and this would require considerable space to perform.†

---

†In this connection I would refer to my work cited above, in which I have specially treated the applications to physics.

(13)   It now remains only to give some indication of the way in which it is possible to distinguish experimentally between the two theories. Before proceeding to this, however, I will mention an experiment which could be regarded as telling against the new theory, but which on more careful consideration is found to lose its value in this connection. According to the new theory unidirectional currents lying in the same straight line can, according to formula (4) exert no influence on each other, while according to the Ampère theory they would repel each other. Now it has been found possible to show that the latter is the case by taking a closed circuit in the form of a rectangle, and making it partially movable in such a way that the movement causes an extension of one pair of sides. Then, without taking account of the other pair of sides, it has been concluded that there must be a repulsive power lying in those parts of the circuit which in this case lie in the same line, and so move apart. To show that this conclusion is incorrect, I need here only refer to the development given above, according to which both theories, when applied to closed circuits, whether or not parts of this circuit are made movable, always lead to the same results. In addition we should also observe for this case that, in the movement of a current from one conductor to another, special forces come into operation which operate in the line of the currents if these lie in the same straight line, and the nature of which is as yet unknown.

(14)   It is quite clear that a decision between the two theories, since the effect of a closed circuit is the same in both cases, can only be made by considering the effect of a limited circuit. But the strength of the current for the same conductor resistance is proportional to the difference in electrical potential at the ends†. If the circuit is a limited one, however, the current cannot proceed further than its boundaries A and B, since otherwise these would not be the end of the circuit. The current will therefore only continue until the potential difference is evened out, and the quantity of electricity passed will be equal to the electrical difference at the two ends. It

---

†This is valid both for the conducting wire of a galvanic circuit, and for a wire attached to a friction machine, except that in the former case the potential difference is always maintained at the same level. From this law it is possible, *a priori*, to arrive at Ohm's law.

follows from this that the maximum effect is obtained when this difference is itself a maximum. The limited circuit must therefore be formed by first charging two spheres as strongly as possible with electricity of opposite sign, and then bringing them into conducting contact (which must only be done after, not during, charging). The effect of this limited circuit on either another electrical circuit, or, better, a magnet, must then be observed, using such an arrangement as will cause the maximum difference between the expected effects on the basis of the two theories.

(15)   Since the incorporation of an amplifier* or a battery, instead of a simple discharge, would cause the limited circuit to approach to a closed circuit, to do this would be to reduce the difference between the effects expected on the basis of the two theories, such amplifying means must not here be used, and it is at once clear with what difficulties experiments of this kind are associated. Since, however, these difficulties are not insuperable, it will be of interest to describe the arrangements which would reveal the greatest difference between the two theories. This occurs, according to my investigations, by using a magnetic needle arranged perpendicular to the rectilinear circuit in such a way that its mid-point lies in the line of the circuit produced, and is capable of free movement at right angles to the plane containing both the circuit and the needle itself. This is shown in Fig. 61 in which AB denotes the limited circuit, so arranged that the positive electricity moves from A to B, and in which two magnets (whose north poles are denoted by N) are connected through an arc SCN made of some solid material hung on to a thread at C.

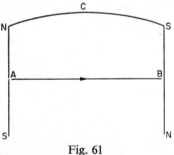

Fig. 61

*Note by R.A.R.T. An amplifier is a galvanometer with many turns of wire.

(16) If we wish to find the effects which in this case the limited circuit would produce according to both theories on an infinitely small magnet, we may do this by replacing the magnet NS by a rectangular current perpendicular to this, lying in the same plane as AB, two of the sides being parallel with AB and the other two perpendicular to this, so arranged that the north pole of the magnet produced by this circuit lies to the left. According to both theories the movement in the direction perpendicular to AB depends now only on the portion of the circuit parallel to AB. If $i.ds$ is a circuit element of AB, while b and b' respectively are the current elements of the rectangular circuit in the same and the opposite directions to AB, then, if r is the distance of the mid-point of this from the middle of the attracting element $i.ds$, the effect on b is repulsive in the perpendicular direction and is equal to $i.ds.b^2/2r^3$.† The same applies to b', though the movement is now attractive. The two effects, since they represent the movement of the rectangular form from b' to b, may be added, and give $i.ds.b^2/r^3$ as the force with which, according to Ampère, the rectangular circuit will be moved in the direction from b' to b. According to my formula the effect on b in the direction perpendicular to this is attractive and equal to $i.ds.b^2/2r^3$, while that on b' is the same in value but repulsive in direction, so that the two operate together to effect movement of the rectangle in the direction from b to b' with force $i.ds.b^2/r^3$. Hence the effects according to the theories are opposite, and this remains true if, instead of the infinitely small circuit element $i.ds$ and an infinitely small magnet, a finite current AB and a finite magnet are substituted, except that in such a case the effects are not of the same magnitude. The effects may be expressed in the following way:

"If in the arrangement given one aligns oneself in the direction of the magnetic needle, with the head to the north end and the feet to the south, and if the eye is then directed to that direction towards which the positive current AB is flowing, then the needle, according to the Ampère theory, will move to the right hand, and, according to the new theory, towards the left hand."

---

†According to equation (1) it is, in the direction of r, equal to $i.ds.b$ $(2—3\cos^2\alpha)/r^2$, and so in the perpendicular direction it is $i.ds.b\ (2—3\cos^2\alpha)$ $\sin \alpha/r^2$, and since when α is infinitely small $\sin \alpha = b/2r$, this gives $— i.ds.b^2/2r^3$, which is thus repulsive.

(17) I will refer in conclusion to two very improbable effects which, according to the Ampère theory, a limited circuit must exert on a magnet. In the first place, a magnet would also be given a rotational movement by a limited circuit, which in the case considered in the previous paragraph would reach its maximum; and, in the second, a magnetic needle which is capable of free rotation about its mid-point would, in the neighbourhood of a limited circuit (if this is the only effect to which it is submitted) not attain to a stable equilibrium but, if it were shifted from the position of equilibrium, turn further either in the one direction or the opposite one, depending on which side it was shifted.

(18) If I may be allowed to hope that the development given makes it seem that the new theory is probably correct in the matters referred to, yet it is still to be desired that experiment should produce a still more decisive distinction between this theory and that due to Ampère. Perhaps some competent physicist will be able to remove all the obstacles which have impeded the experiments to which I have referred.

Index